DESCAR.

Antonio Damasio is David Dornsife Professor of
Neuroscience, Neurology and Psychology at the
University of Southern California, where he directs
the new Brain and Creativity Institute. He is also
adjunct professor at the Salk Institute and at
the University of Iowa. He is the recipient of
numerous awards (several shared with his wife
Hanna Damasio, also a neurologist and neuro-
scientist), most recently the 2005 Prince of Asturias
Prize in Science and Technology. He is a member of
the Institute of Medicine of the National Academy
of Sciences, a fellow of the American Academy of
Arts and Sciences, and the author of two other
widely acclaimed books, *The Feeling of What
Happens* and *Looking for Spinoza*, which have
been translated into over 30 languages.

'Damasio has written his book with the literary skill of a suspense novel and yet it offers sound, easily accessible and reliable information about what is known of the anatomy, organization and functions of the forebrain. Educated laymen curious about human biology, medical students, neurologists, other physicians and surgeons, sociologists, psychologists and anthropologists should, by all means, read this book.'

Integrative Physiological and Behavioural Science

'Damasio lays out a provocative theory...emotion is part and parcel of what we call cognition. If there is severe impairment of the emotions, we cannot have rationality.'

Washington Post

'Here at last is an attempt, by one of the world's foremost neurologists, to synthesize what is known about the workings of the human brain. It bases its arguments on a profound knowledge of the brain, rather than on a wish to redesign it as an engineer might. It deserves to become a classic.'

David Hubel, Nobel Laureate, Harvard University

'Antonio Damasio's astonishing book takes us on a scientific journey into the brain that reveals the invisible world within as if it were visible to our sight. You will never again look at yourself or at another without wondering what goes on behind the eyes that so meet.'

Jonas Salk, Biologist

'Damasio has written an engaging, informative book that challenges the dogma that emotions interfere with wise decisions, and that places feelings in their proper role in human functioning. David Hume should be smiling.'

Jerome Kagan, Daniel and Amy Starch Professor of Psychology, Harvard University

ANTONIO DAMASIO

Descartes' Error

Emotion, Reason and the Human Brain

VINTAGE BOOKS
London

Published by Vintage 2006

10 9

First published in 1994 in the United States of America by
G.P. Putnam's Sons, New York

First published in Great Britain in 1995 by Picador

Vintage
Random House, 20 Vauxhall Bridge Road, London SW1V 2SA

www.vintage-books.co.uk

Addresses for companies within The Random House Group Limited can be
found at: www.randomhouse.co.uk/offices.htm

The Random House Group Limited Reg. No. 954009

A CIP catalogue record for this book is available from the British Library

All figures are original. The figure on page 28 was prepared by Kathleen
Rockland. All others are by Hanna Damasio. The figure on page 104 contains
a photomicrograph from the work of Roger Tootell, reproduced with his
permission and that of the *Journal of Neuroscience*. The figures on pages 141
and 143 contain photos of Julie Fiez, used with her permission.

ISBN 9780099501640

The Random House Group Limited supports The Forest
Stewardship Council (FSC), the leading international forest
certification organisation. All our titles that are printed on
Greenpeace approved FSC certified paper carry the FSC logo.
Our paper procurement policy can be found at
www.rbooks.co.uk/environment.

Mixed Sources
Product group from well-managed
forests and other controlled sources
www.fsc.org   Cert no. TT-COC-2139
© 1996 Forest Stewardship Council
FSC

Printed in the UK by CPI Bookmarque, Croydon, CR0 4TD

FOR HANNA

Contents

PART III

Preface

If we were alive around 1900, and were in any way interested in intellectual matters, we probably would have thought that the time had come for science to tackle the understanding of emotion in its many dimensions and answer the public's growing curiosity about it in a definitive way. In the preceding decades Charles Darwin had shown how some emotional phenomena are present in remarkably comparable ways in nonhuman species; William James and Carl Lange had advanced an innovative proposal to explain the processing of emotions; Sigmund Freud had turned the emotions into the centerpiece of his inquiry on psychopathological states; and Charles Sherrington had begun the neurophysiological investigation of the brain circuits involved in emotion. Nonetheless, the all-out attack on the subject of emotion, there and then, never came to pass. On the contrary, as the sciences of mind and brain flourished in the twentieth century, interests went elsewhere and the specialties which we loosely group today under neuroscience gave a resolute cold shoulder to emotion research. True, the psychoanalysts never forgot the emotions, and there were noble exceptions—pharmacologists and psychiatrists concerned with disorders of mood, and lone psychologists and neuroscientists who cultivated an interest in affect. Those exceptions, however, merely accentuated the neglect of emotion as a research topic. Behaviorism, the cognitive revolution, and computational neuroscience did not reduce this neglect in any appreciable way.

By and large this was still the state of affairs in 1994 when *Descartes' Error* was first published, although the ground had already begun to shift. The book was, through and through, about the brain science of emotion and about its implications for decision-making in general and for social behavior in particular. I hoped to make my point quietly without being thrown off the stage but I had no right to expect welcome signs and an attentive audience. But I did get a welcoming, attentive, and generous audience, here and abroad, and a number of the ideas in the book have found their way into the thinking of many colleagues and of the nonspecialist public. Just as unexpected was the fact that so many readers were eager to engage in a conversation, pose questions, make suggestions, and offer corrections. In several instances I corresponded with those readers, some of whom have become friends. I learned a lot, and I still do, since hardly a day goes by without mail about *Descartes' Error* from somewhere in the world.

A decade later the situation is radically different. Not long after *Descartes' Error*, two of the neuroscientists who had been studying emotions in animals published their own books: *The Emotional Brain* (1996) by Joseph Le Doux and *Affective Neuroscience* (1998) by Jaak Panksepp. Others followed and soon neuroscience laboratories, in America and in Europe, had turned their attention to emotion research. Philosophers cultivating the subject were heard with a new attention. (Martha Nussbaum was a particularly good example of this.) And books capitalizing on the science of emotion became widely popular (Daniel Goleman's *Emotional Intelligence*, for example). Emotion is finally being given the due that our illustrious forerunners would have wished it to receive, albeit a century late.

The main subject in *Descartes' Error* is the relation between emotion and reason. Based on my study of neurological patients who had both defects of decision-making and a disorder of emotion, I advanced the hypothesis (known as the somatic marker hypothesis) that emotion was in the loop of reason, and that emotion could assist the reasoning process rather than necessarily disturb it, as was commonly assumed. Today this idea does not cause any raised eyebrows although at the time I presented the notion it startled many and was even regarded with some skepticism. On balance, the idea was largely embraced, so embraced that, on occasion, it was bent out of shape. For example, I never wrote, as was later suggested, that the assistance emotion provides to reasoning would necessarily occur nonconsciously. On the contrary, my first proposal equated somatic markers with conscious gut feelings, although I did make room for a nonconscious variety of somatic marker; nor did I regard skin conductance responses as somatic markers, but rather as indices of somatic markers. Finally, I never suggested that emotion was a substitute for reasoning, but in some superficial versions of the work it sounded as if I was proposing that if you follow your heart instead of your reason all would be well.

To be sure, on certain occasions, emotions can be a substitute for reason. The emotional action program we call fear can get most human beings out of danger, in short order, with little or no help from reason. A squirrel or a bird will respond to a threat without any thinking at all and the same can happen to a human. In effect, in some circumstances, too much thinking may be far less advantageous than no thinking at all. That is the beauty of how emotion has functioned throughout evolution: it allows the possibility of making living beings *act* smartly without having to *think* smartly. In humans, however, this story has become more complicated, for better and for worse. Reasoning does what emotions do but achieves it knowingly. Reasoning gives us the option of thinking smartly *before* we act smart, and a good thing too: we have discovered that the emotions alone can solve many—but not all—the problems posed by our complex environment and that, on occasion, the solutions offered by emotion are actually counterproductive.

But how did the complex species evolve the smart reasoning system? The new proposal in *Descartes' Error* is that the reasoning system evolved as an extension of the automatic emotional system, with emotion playing diverse roles in the reasoning process. For example, emotion may increase the saliency of a premise and, in so doing, bias the conclusion in favor of the premise. Emotion also assists with the process of holding in mind the multiple facts that must be considered in order to reach a decision.

The obligate participation of emotion in the reasoning process can be advantageous or nefarious depending both on the circumstances of the decision and on the past history of the decider. The issue of circumstances is well illustrated by the story with which Malcolm Gladwell opens his book *Blink* (2005). The curators of the Getty Museum concluded that a certain Greek sculpture was the real thing in the context of their desire to add the piece to the collection. A number of external experts, on the other hand, judged the piece to be a fake based on their gut feeling of rejection upon seeing it for the first time. Emotions of different kinds participated in these two different judgments at different stages of the reasoning process. There was a rewarding and pervasive desire to endorse the object for some; and there was the immediately punitive and thoroughly conscious gut feeling that something was amiss for others. In neither case, however, did reason operate alone, and that is the critical point I made in *Descartes' Error*. When emotion is entirely left out of the reasoning picture, as happens in certain neurological conditions, reason turns out to be even more flawed than when emotion plays bad tricks on our decisions.

The somatic marker hypothesis postulated from its inception that emotions *marked* certain aspects of a situation, or certain outcomes of possible actions. Emotion achieved this marking quite overtly, as in a "gut feeling," or covertly, via signals occurring below the radar of our awareness (examples of covert signals would be neuromodulator responses, such as those of dopamine or oxytocin, which can change the behavior of neuron groups that represent a certain choice). As for the knowledge used in reasoning, it too could be fairly explicit or partially hidden, as when we intuit a solution. In other words, emotion had a role to play in intuition, the sort of rapid cognitive process in which we come to a particular conclusion without being aware of all the immediate logical steps. It is not necessarily the case that the knowledge of the intermediate steps is absent, only that emotion delivers the conclusion so directly and rapidly that not much knowledge need come to mind. This is in keeping with the old saying which tells us that "intuition favors the prepared mind." What does the saying mean in the context of the somatic marker hypothesis? The quality of one's intuition depends on how well we have reasoned in the past; on how well we have classified the events of our past experience in relation to the emotions that preceded and followed them; and also on how well we have reflected on the successes and failures of our past intuitions. Intuition is simply rapid cognition with the required knowledge partially swept under the carpet, all courtesy of emotion and much past practice. Clearly I never wished to set emotion against reason, but rather to see emotion as at least assisting reason and at best holding a dialogue with it. Nor did I ever oppose emotion to cognition since I view emotion as delivering cognitive information, directly and via feelings.

The evidence that formed the basis for the somatic marker hypothesis emerged over several years from the study of neurological patients whose social conduct had been altered by brain damage in a specific sector of the frontal lobe. The observations in those patients eventually led to another important idea in *Descartes' Error*: the notion that the brain systems that are jointly engaged in emotion and decision-making are generally involved in the management of social cognition and behavior. This notion opened the way for connecting the fabric of social and cultural phenomena to specific features of neurobiology, a connection supported by powerful facts.

The publication of *Descartes' Error* is responsible for a related discovery. Parents of young men and women who resembled our adult-onset frontal patients in some aspects of their social behavior wrote to me wondering, quite perceptively, whether the troubles of their now grown up children might be due to brain damage too. We found out that it was, as reported in the very first studies on this issue, which were published in 1999. These young adults had suffered frontal brain damage early in their lives, a fact that had either not been known to the parents or had not been connected with their manifestly abnormal social behavior. We also discovered a fundamental way in which the early-onset cases differed from the adult-onset cases: the early-onset patients appeared not to have learned the social conventions and ethical rules that should have governed their behavior. Whereas the adult-onset patients knew the rules but failed to act according to them, the early-onset case had never learned the rules to begin with. In other words, while the adult-onset cases told us that emotions were required for the deployment of proper social behavior; the early-onset cases showed that emotions were also needed for mastering the know-how behind proper social behavior. The implications of this fact for understanding the possible causes of disordered social conduct are barely beginning to be appreciated.

The postscriptum of *Descartes' Error* contained an idea which pointed to the future of neurobiological research: the mechanisms of basic homeostasis constitute a blueprint for the cultural development of the human values which permit us to judge actions as good or evil, and classify objects as beautiful or ugly. At the time, writing about this idea gave me hope that a two-way bridge could be established between neurobiology and the humanities, thus providing the way for a better understanding of human conflict and for a more comprehensive account of creativity. I am pleased to report that some progress has been made toward building that sort of bridge. For example, some of us are actively investigating the brain states associated with moral reasoning while others are trying to discover what the brain does during esthetic experiences. The intent is not to reduce ethics or esthetics to brain circuitry but rather to explore the threads that interconnect neurobiology to culture. I am even more hopeful today that such a seemingly utopian bridge can become reality and optimistic that we will enjoy its benefits without having to wait another century.

Antonio Damasio, 2005

Introduction

ALTHOUGH I CANNOT tell for certain what sparked my interest in the neural underpinnings of reason, I do know when I became convinced that the traditional views on the nature of rationality could not be correct. I had been advised early in life that sound decisions came from a cool head, that emotions and reason did not mix any more than oil and water. I had grown up accustomed to thinking that the mechanisms of reason existed in a separate province of the mind, where emotion should not be allowed to intrude, and when I thought of the brain behind that mind, I envisioned separate neural systems for reason and emotion. This was a widely held view of the relation between reason and emotion, in mental and neural terms.

But now I had before my eyes the coolest, least emotional, intelligent human being one might imagine, and yet his practical reason was so impaired that it produced, in the wanderings of daily life, a succession of mistakes, a perpetual violation of what would be considered socially appropriate and personally advantageous. He had had an entirely healthy mind until a neurological disease ravaged a specific sector of his brain and, from one day to the next, caused this profound defect in decision making. The instruments usually considered necessary and sufficient for rational behavior were intact in him. He had the requisite knowledge, attention, and memory; his language was flawless; he could perform calculations; he could tackle the logic of an abstract problem. There was only one significant accompaniment to his decision-making failure: a marked alteration of the ability to experience feelings. Flawed reason and impaired feelings stood out together as the consequences of a specific brain lesion, and this correlation suggested to me that feeling was an integral component of the machinery of reason. Two decades of clinical and experimental work with a large number of neurological patients have allowed me to replicate this observation many times, and to turn a clue into a testable hypothesis.[1]

I began writing this book to propose that reason may not be as pure as most of us think it is or wish it were, that emotions and feelings may not be intruders in the bastion of reason at all: they may be enmeshed in its networks, for worse *and* for better. The strategies of human reason probably did not develop, in either evolution or any single individual, without the guiding force of the mechanisms of biological regulation, of which emotion and feeling are notable expressions. Moreover, even after reasoning strategies become established in the formative years, their effective deployment probably depends, to a considerable extent, on a continued ability to experience feelings.

This is not to deny that emotions and feelings can cause havoc in the processes of reasoning under certain circumstances. Traditional wisdom has told us that they can, and recent investigations of the normal reasoning process also reveal the potentially harmful influence of emotional biases. It is thus even more surprising and novel that the *absence* of emotion and feeling is no less damaging, no less capable of compromising the rationality that makes us distinctively human and allows us to decide in consonance with a sense of personal future, social convention, and moral principle.

Nor is this to say that when feelings have a positive action they do the deciding for us; or that we are not rational beings. I suggest only that certain aspects of the process of emotion and feeling are indispensable for rationality. At their best, feelings point us in the proper direction, take us to the appropriate place in a decision-making space, where we may put the instruments of logic to good use. We are faced by uncertainty when we have to make a moral judgment, decide on the course of a personal relationship, choose some means to prevent our being penniless in old age, or plan for the life that lies ahead. Emotion and feeling, along with the covert physiological machinery underlying them, assist us with the daunting task of predicting an uncertain future and planning our actions accordingly.

Beginning with an analysis of the nineteenth-century landmark case of Phineas Gage, whose behavior first revealed a connection between impaired rationality and specific brain damage, I examine recent investigations of his modern counterparts and review pertinent findings from neuropsychological research in humans and animals. Further, I propose that human reason depends on several brain systems, working in concert across many levels of neuronal organization, rather than on a single brain center. Both "high-level" and "low-level" brain regions, from the prefrontal cortices to the hypothalamus and brain stem, cooperate in the making of reason.

The lower levels in the neural edifice of reason are the same ones that regulate the processing of emotions and feelings, along with the body functions necessary for an organism's survival. In turn, these lower levels maintain direct and mutual relationships with virtually every bodily organ, thus placing the body directly within the chain of operations that generate the highest reaches of reasoning, decision making, and, by extension, social behavior and creativity. Emotion, feeling, and biological regulation all play a role in human reason. The lowly orders of our organism are in the loop of high reason.

It is intriguing to find the shadow of our evolutionary past at the most distinctively human level of mental function, although Charles Darwin prefigured the essence of this finding when he wrote about the indelible stamp of lowly origins which humans bear in their bodily frame.[2] Yet the dependence of high reason on low brain does not turn high reason into low reason. The fact that acting according to an ethical principle requires the participation of simple circuitry in the brain core does not cheapen the ethical principle. The edifice of ethics does not collapse, morality is not threatened, and in a normal individual the will remains the will. What can change is our view of how biology has contributed to the origin of certain ethical principles arising in a social context, when many individuals with a similar biological disposition interact in specific circumstances.

Feeling is the second and central topic of this book, and one to which I was drawn not by design but by necessity, as I struggled to understand the cognitive and neural machinery behind reasoning and decision making. A second idea in the book, then, is that the essence of a feeling may not be an elusive mental quality attached to an object, but rather the direct perception of a specific landscape: that of the body.

My investigation of neurological patients in whom brain lesions impaired the experience of feelings has led me to think that feelings are not as intangible as they have been presumed to be. One may be able to pin them down mentally, and perhaps find their neural substrate as well. In a departure from current neurobiological think-ing, I propose that the critical networks on which feelings rely include not only the traditionally acknowledged collection of brain structures known as the limbic system but also some of the brain's prefrontal cortices, and, most importantly, the brain sectors that map and integrate signals from the body.

I conceptualize the essence of feelings as something you and I can see through a window that opens directly onto a continuously up-dated image of the structure and state of our body. If you imagine the view from this window as a landscape, the body "structure" is analo-gous to object shapes in a space, while the body "state" resembles the light and shadow and movement and sound of the objects in that space. In the landscape of your body, the objects are the viscera (heart, lungs, gut, muscles), while the light and shadow and move-ment and sound represent a point in the range of operation of those organs at a certain moment. By and large, a feeling is the momentary "view" of a part of that body landscape. It has a specific content—the state of the body; and specific neural systems that support it—the peripheral nervous system and the brain regions that integrate signals related to body structure and regulation. Because the sense of that body landscape is juxtaposed in time to the perception or recollection of something else that is not part of the body—a face, a melody, an aroma—feelings end up being "qualifiers" to that something else. But there is more to a feeling than this essence. As I will explain, the qualifying body state, positive or negative, is accompanied and rounded up by a corresponding thinking mode: fast moving and idea rich, when the body-state is in the positive and pleasant band of the spectrum, slow moving and repetitive, when the body-state veers toward the painful band.

In this perspective, feelings are the sensors for the match or lack thereof between nature and circumstance. And by nature I mean both the nature we inherited as a pack of genetically engineered adaptations, and the nature we have acquired in individual development, through interactions with our social environment, mindfully and willfully as well as not. Feelings, along with the emotions they come from, are not a luxury. They serve as internal guides, and they help us communicate to others signals that can also guide them. And feelings are neither intangible nor elusive. Contrary to traditional scientific opinion, feelings are just as cognitive as other percepts. They are the result of a most curious physiological arrangement that has turned the brain into the body's captive audience.

Feelings let us catch a glimpse of the organism in full biological swing, a reflection of the mechanisms of life itself as they go about their business. Were it not for the possibility of sensing body states that are inherently ordained to be painful or pleasurable, there would be no suffering or bliss, no longing or mercy, no tragedy or glory in the human condition.

At first glance, the view of the human spirit proposed here may not be intuitive or comforting. In attempting to shed light on the complex phenomena of the human mind, we run the risk of merely degrading them and explaining them away. But that will happen only if we confuse a phenomenon itself with the separate components and operations that can be found behind its appearance. I am not suggesting that.

To discover that a particular feeling depends on activity in a number of specific brain systems interacting with a number of body organs does not diminish the status of that feeling as a human phenomenon. Neither anguish nor the elation that love or art can bring about are devalued by understanding some of the myriad biological processes that make them what they are. Precisely the opposite should be true: Our sense of wonder should increase before the intricate mechanisms that make such magic possible. Feelings form the base for what humans have described for millennia as the human soul or spirit.

This book is also about a third and related topic: that the body, as represented in the brain, may constitute the indispensable frame of reference for the neural processes that we experience as the mind; that our very organism rather than some absolute external reality is used as the ground reference for the constructions we make of the world around us and for the construction of the ever-present sense of subjectivity that is part and parcel of our experiences; that our most refined thoughts and best actions, our greatest joys and deepest sorrows, use the body as a yardstick.

Surprising as it may sound, the mind exists in and for an integrated organism; our minds would not be the way they are if it were not for the interplay of body and brain during evolution, during individual development, and at the current moment. The mind had to be first about the body, or it could not have been. On the basis of the ground reference that the body continuously provides, the mind can then be about many other things, real and imaginary.

This idea is anchored in the following statements: (1) The human brain and the rest of the body constitute an indissociable organism, integrated by means of mutually interactive biochemical and neural regulatory circuits (including endocrine, immune, and autonomic neural components); (2) The organism interacts with the environment as an ensemble: the interaction is neither of the body alone nor of the brain alone; (3) The physiological operations that we call mind are derived from the structural and functional ensemble rather than from the brain alone: mental phenomena can be fully understood only in the context of an organism's interacting in an environment. That the environment is, in part, a product of the organism's activity itself, merely underscores the complexity of interactions we must take into account.

It is not customary to refer to organisms when we talk about brain and mind. It has been so obvious that mind arises from the activity of neurons that only neurons are discussed as if their operation could be independent from that of the rest of the organism. But as I investigated disorders of memory, language, and reason in numerous human beings with brain damage, the idea that mental activity, from its simplest aspects to its most sublime, requires both brain and body proper became especially compelling. I believe that, relative to the brain, the body proper provides more than mere support and modulation: it provides a basic topic for brain representations.

There are facts to support this idea, reasons why the idea is plausible, and reasons why it would be nice if things really were this way. Foremost among the last is that the body precedence proposed here might shed light on one of the most vexing of all questions since humans began inquiring about their minds: How is it that we are conscious of the world around us, that we know what we know, and that we know that we know?

In the perspective of the above hypothesis, love and hate and anguish, the qualities of kindness and cruelty, the planned solution of a scientific problem or the creation of a new artifact are all based on neural events within a brain, provided that brain has been and now is interacting with its body. The soul breathes through the body, and suffering, whether it starts in the skin or in a mental image, happens in the flesh.

. . .

I wrote this book as my side of a conversation with a curious, intelligent, and wise imaginary friend, who knew little about neuroscience but much about life. We made a deal: the conversation was to have mutual benefits. My friend was to learn about the brain and about those mysterious things mental, and I was to gain insights as I struggled to explain my idea of what body, brain, and mind are about. We agreed not to turn the conversation into a boring lecture, not to disagree violently, and not to try to cover too much. I would talk about established facts, about facts in doubt, and about hypotheses, even when I could come up with nothing but hunches to support them. I would talk about work in progress literally, about several research projects then under way, and about work that would start long after the conversation was over. It was also understood that, as befits a conversation, there would be byways and diversions, as well as passages that would not be clear the first time around and might benefit from a second visit. That is why you will find me returning to some topics, every now and then, from a different perspective.

At the outset I made my view clear on the limits of science: I am skeptical of science's presumption of objectivity and definitiveness. I have a difficult time seeing scientific results, especially in neurobiology, as anything but provisional approximations, to be enjoyed for a while and discarded as soon as better accounts become available. But skepticism about the current reach of science, especially as it concerns the mind, does not imply diminished enthusiasm for the attempt to improve provisional approximations.

Perhaps the complexity of the human mind is such that the solution to the problem can never be known because of our inherent limitations. Perhaps we should not even talk about a problem at all, and speak instead of a mystery, drawing on a distinction between questions that can be approached suitably by science and questions that are likely to elude science forever.[3] But much as I have sympathy for those who cannot imagine how we might unravel the mystery (they have been dubbed "mysterians"[4]), and for those who think it is knowable but would be disappointed if the explanation were to rely on something already known, I do believe, more often than not, that we will come to know.

By now you may have concluded that the conversation was neither about Descartes nor about philosophy, although it certainly was about mind, brain, and body. My friend suggested it should take place under the Sign of Descartes, since there was no way of approaching such themes without evoking the emblematic figure who shaped the most commonly held account of their relationship. At this point I realized that, in a curious way, the book would be about Descartes' Error. You will, of course, want to know what the Error was, but for the moment I am sworn to secrecy. I promise, though, that it will be revealed.

Our conversation then began in earnest, with the strange life and times of Phineas Gage.

Part
_____
1

One

Unpleasantness in Vermont

PHINEAS P. GAGE

IT IS THE summer of 1848. We are in New England. Phineas P. Gage, twenty-five years old, construction foreman, is about to go from riches to rags. A century and a half later his downfall will still be quite meaningful.

Gage works for the Rutland & Burlington Railroad and is in charge of a large group of men, a "gang" as it is called, whose job it is to lay down the new tracks for the railroad's expansion across Vermont. Over the past two weeks the men have worked their way slowly toward the town of Cavendish; they are now at a bank of the Black River. The assignment is anything but easy because of the outcrops of hard rock. Rather than twist and turn the tracks around every escarpment, the strategy is to blast the stone and make way for a straighter and more level path. Gage oversees these tasks and is equal to them in every way. He is five-foot-six and athletic, and his movements are swift and precise. He looks like a young Jimmy Cagney, a Yankee Doodle dandy dancing his tap shoes over ties and tracks, moving with vigor and grace.

In the eyes of his bosses, however, Gage is more than just another able body. They say he is "the most efficient and capable" man in their employ.[1] This is a good thing, because the job takes as much physical prowess as keen concentration, especially when it comes to preparing the detonations. Several steps have to be followed, in orderly fashion. First, a hole must be drilled in the rock. After it is filled about halfway with explosive powder, a fuse must be inserted, and the powder covered with sand. Then the sand must be "tamped in," or pounded with a careful sequence of strokes from an iron rod. Finally, the fuse must be lit. If all goes well, the powder will explode into the rock; the sand is essential, for without its protection the explosion would be directed away from the rock. The shape of the iron and the way it is played are also important. Gage, who has had an iron manufactured to his specifications, is a virtuoso of this thing.

Now for what is going to happen. It is four-thirty on this hot afternoon. Gage has just put powder and fuse in a hole and told the man who is helping him to cover it with sand. Someone calls from behind, and Gage looks away, over his right shoulder, for only an instant. Distracted, and before his man has poured the sand in, Gage begins tamping the powder directly with the iron bar. In no time he strikes fire in the rock, and the charge blows upward in his face.[2]

The explosion is so brutal that the entire gang freezes on their feet. It takes a few seconds to piece together what is going on. The bang is unusual, and the rock is intact. Also unusual is the whistling sound, as of a rocket hurled at the sky. But this is more than fireworks. It is assault and battery. The iron enters Gage's left cheek, pierces the base of the skull, traverses the front of his brain, and exits at high speed through the top of the head. The rod has landed more than a hundred feet away, covered in blood and brains. Phineas Gage has been thrown to the ground. He is stunned, in the afternoon glow, silent but awake. So are we all, helpless spectators.

"Horrible Accident" will be the predictable headline in the Boston *Daily Courier* and *Daily Journal* of September 20, a week later.

"Wonderful Accident" will be the strange headline in the *Vermont Mercury* of September 22. "Passage of an Iron Rod Through the Head" will be the accurate headline in the *Boston Medical and Surgical Journal*. From the matter-of-factness with which they tell the story, one would think the writers were familiar with Edgar Allan Poe's accounts of the bizarre and the horrific. And perhaps they were, although this is not likely; Poe's gothic tales are not yet popular, and Poe himself will die the next year, unknown and impecunious. Perhaps the horrible is just in the air.

Noting how surprised people were that Gage was not killed instantly, the Boston medical article documents that "immediately after the explosion the patient was thrown upon his back"; that shortly thereafter he exhibited "a few convulsive motions of the extremities," and "spoke in a few minutes"; that "his men (with whom he was a great favourite) took him in their arms and carried him to the road, only a few rods distant (a rod is equivalent to 5½ yards, or 16½ feet), and sat him into an ox cart, in which he rode, sitting erect, a full three quarters of a mile, to the hotel of Mr. Joseph Adams"; and that Gage "got out of the cart himself, with a little assistance from his men."

Let me introduce Mr. Adams. He is the justice of the peace for Cavendish and the owner of the town's hotel and tavern. He is taller than Gage, twice as round, and as solicitous as his Falstaff shape suggests. He approaches Gage, and immediately has someone call for Dr. John Harlow, one of the town physicians. While they wait, I imagine, he says, "Come, come, Mr. Gage, what have we got here?" and, why not, "My, my, what troubles we've seen." He shakes his head in disbelief and leads Gage to the shady part of the hotel porch, which has been described as a "piazza." That makes it sound grand and spacious and open, and perhaps it is grand and spacious, but it is not open; it is just a porch. And there perhaps Mr. Adams is now giving Phineas Gage lemonade, or maybe cold cider.

An hour has passed since the explosion. The sun is declining and the heat is more bearable. A younger colleague of Dr. Harlow's, Dr. Edward Williams, is arriving. Years later Dr. Williams will describe the scene: "He at that time was sitting in a chair upon the piazza of Mr. Adams' hotel, in Cavendish. When I drove up, he said, 'Doctor, here is business enough for you.' I first noticed the wound upon the head before I alighted from my carriage, the pulsations of the brain being very distinct; there was also an appearance which, before I examined the head, I could not account for: the top of the head appeared somewhat like an inverted funnel; this was owing, I discovered, to the bone being fractured about the opening for a distance of about two inches in every direction. I ought to have mentioned above that the opening through the skull and integuments was not far from one and a half inch in diameter; the edges of this opening were everted, and the whole wound appeared as if some wedge-shaped body had passed from below upward. Mr. Gage, during the time I was examining this wound, was relating the manner in which he was injured to the bystanders; he talked so rationally and was so willing to answer questions, that I directed my inquiries to him in preference to the men who were with him at the time of the accident, and who were standing about at this time. Mr. G. then related to me some of the circumstances, as he has since done; and I can safely say that neither at that time nor on any subsequent occasion, save once, did I consider him to be other than perfectly rational. The one time to which I allude was about a fortnight after the accident, and then he persisted in calling me John Kirwin; yet he answered all my questions correctly."[3]

The survival is made all the more amazing when one considers the shape and weight of the iron bar. Henry Bigelow, a surgery professor at Harvard, describes the iron so: "The iron which thus traversed the skull weighs thirteen and a quarter pounds. It is three feet seven inches in length, and one and a quarter inches in diameter. The end which entered first is pointed; the taper being seven inches long, and the diameter of the point one quarter of an inch; circumstances to which the patient perhaps owes his life. The iron is unlike any other, and was made by a neighbouring blacksmith to please the fancy of the owner."[4] Gage is serious about his trade and its proper tools.

Surviving the explosion with so large a wound to the head, being able to talk and walk and remain coherent immediately afterward—this is all surprising. But just as surprising will be Gage's surviving the inevitable infection that is about to take over his wound. Gage's physician, John Harlow, is well aware of the role of disinfection. He does not have the help of antibiotics, but using what chemicals are available he will clean the wound vigorously and regularly, and place the patient in a semi-recumbent position so that drainage will be natural and easy. Gage will develop high fevers and at least one abscess, which Harlow will promptly remove with his scalpel. In the end, Gage's youth and strong constitution will overcome the odds against him, assisted, as Harlow will put it, by divine intervention: "I dressed him, God healed him."

Phineas Gage will be pronounced cured in less than two months. Yet this astonishing outcome pales in comparison with the extraordinary turn that Gage's personality is about to undergo. Gage's disposition, his likes and dislikes, his dreams and aspirations are all to change. Gage's body may be alive and well, but there is a new spirit animating it.

GAGE WAS NO LONGER GAGE

Just what exactly happened we can glean today from the account Dr. Harlow prepared twenty years after the accident.[5] It is a trustworthy text, with an abundance of facts and a minimum of interpretation. It makes sense humanly and neurologically, and from it we can piece together not just Gage but his doctor as well. John Harlow had been a schoolteacher before he entered Jefferson Medical College in Philadelphia, and was only a few years into his medical career when he took care of Gage. The case became his life-consuming interest, and I suspect that it made Harlow want to be a scholar, something that may not have been in his plans when he set up his medical practice in Vermont. Treating Gage successfully and reporting the results to his Boston colleagues may have been the shining hours of his career, and he must have been disturbed by the fact that a real cloud hung over Gage's cure.

Harlow's narrative describes how Gage regained his strength and how his physical recovery was complete. Gage could touch, hear, and see, and was not paralyzed of limb or tongue. He had lost vision in his left eye, but his vision was perfect in the right. He walked firmly, used his hands with dexterity, and had no noticeable difficulty with speech or language. And yet, as Harlow recounts, the "equilibrium or balance, so to speak, between his intellectual faculty and animal propensities" had been destroyed. The changes became apparent as soon as the acute phase of brain injury subsided. He was now "fitful, irreverent, indulging at times in the grossest profanity which was not previously his custom, manifesting but little deference for his fellows, impatient of restraint or advice when it conflicts with his desires, at times pertinaciously obstinate, yet capricious and vacillating, devising many plans of future operation, which are no sooner arranged than they are abandoned. . . . A child in his intellectual capacity and manifestations, he has the animal passions of a strong man." The foul language was so debased that women were advised not to stay long in his presence, lest their sensibilities be offended. The strongest admonitions from Harlow himself failed to return our survivor to good behavior.

These new personality traits contrasted sharply with the "temperate habits" and "considerable energy of character" Phineas Gage was known to have possessed before the accident. He had had "a well balanced mind and was looked upon by those who knew him as a shrewd, smart businessman, very energetic and persistent in executing all his plans of action." There is no doubt that in the context of his job and time, he was successful. So radical was the change in him that friends and acquaintances could hardly recognize the man. They noted sadly that "Gage was no longer Gage." So different a man was he that his employers would not take him back when he returned to work, for they "considered the change in his mind so marked that they could not give him his place again." The problem was not lack of physical ability or skill; it was his new character.

The unraveling continued unabated. No longer able to work as a foreman, Gage took jobs on horse farms. One gathers that he was prone to quit in a capricious fit or be let go because of poor discipline. As Harlow notes, he was good at "always finding something which did not suit him." Then came his career as a circus attraction. Gage was featured at Barnum's Museum in New York City, vaingloriously showing his wounds and the tamping iron. (Harlow states that the iron was a constant companion, and points out Gage's strong attachment to objects and animals, which was new and somewhat out of the ordinary. This trait, what we might call "collector's behavior," is something I have seen in patients who have suffered injuries like Gage's, as well as in autistic individuals.)

Then far more than now, the circus capitalized on nature's cruelty. The endocrine variety included dwarfs, the fattest woman on earth, the tallest man, the fellow with the largest jaw; the neurological variety included youths with elephant skin, victims of neurofibromatosis—and now Gage. We can imagine him in such company, peddling misery for gold.

Four years after the accident, there was another theatrical coup. Gage left for South America. He may have worked on horse farms, and was a sometime stagecoach driver in Santiago and Valparaiso. Little else is known about his expatriate life except that in 1859 his health was deteriorating.

In 1860, Gage returned to the United States to live with his mother and sister, who had since moved to San Francisco. At first he was employed on a farm in Santa Clara, but he did not stay long. In fact, he moved around, occasionally finding work as a laborer in the area. It is clear that he was not an independent person and that he could not secure the type of steady, remunerative job that he had once held. The end of the fall was nearing.

In my mind is a picture of 1860s San Francisco as a bustling place, full of adventurous entrepreneurs engaged in mining, farming, and shipping. That is where we can find Gage's mother and sister, the latter married to a prosperous San Francisco merchant (D. D. Shattuck, Esquire), and that is where the old Phineas Gage might have belonged. But that is not where we would find him if we could travel back in time. We would probably find him drinking and brawling in a questionable district, not conversing with the captains of commerce, as astonished as anybody when the fault would slip and the earth would shake threateningly. He had joined the tableau of dispirited people who, as Nathanael West would put it decades later, and a few hundred miles to the south, "had come to California to die."[6]

The meager documents available suggest that Gage developed epileptic fits (seizures). The end came on May 21, 1861, after an illness that lasted little more than a day. Gage had a major convulsion which made him lose consciousness. A series of subsequent convulsions, one coming soon on the heels of another, followed. He never regained consciousness. I believe he was the victim of *status epilepticus*, a condition in which convulsions become nearly continuous and usher in death. He was thirty-eight years old. There was no death notice in the San Francisco newspapers.

WHY PHINEAS GAGE?

Why is this sad story worth telling? What is the possible significance of such a bizarre tale? The answer is simple. While other cases of neurological damage that occurred at about the same time revealed that the brain was the foundation for language, perception, and motor function, and generally provided more conclusive details, Gage's story hinted at an amazing fact: Somehow, there were systems in the human brain dedicated more to reasoning than to anything else, and in particular to the personal and social dimensions of reasoning. The observance of previously acquired social convention and ethical rules could be lost as a result of brain damage, even when neither basic intellect nor language seemed compromised. Unwittingly, Gage's example indicated that something in the brain was concerned specifically with unique human properties, among them the ability to anticipate the future and plan accordingly within a complex social environment; the sense of responsibility toward the self and others; and the ability to orchestrate one's survival deliberately, at the command of one's free will.

The most striking aspect of this unpleasant story is the discrepancy between the normal personality structure that preceded the accident and the nefarious personality traits that surfaced thereafter and seem to have remained for the rest of Gage's life. Gage had once known all he needed to know about making choices conducive to his betterment. He had a sense of personal and social responsibility, reflected in the way he had secured advancement in his job, cared for the quality of his work, and attracted the admiration of employers and colleagues. He was well adapted in terms of social convention and appears to have been ethical in his dealings. After the accident, he no longer showed respect for social convention; ethics in the broad sense of the term, were violated; the decisions he made did not take into account his best interest, and he was given to invent tales "without any foundation except in his fancy," in Harlow's words. There was no evidence of concern about his future, no sign of forethought.

The alterations in Gage's personality were not subtle. He could not make good choices, and the choices he made were not simply neutral. They were not the reserved or slight decisions of someone whose mind is diminished and who is afraid to act, but were instead actively disadvantageous. One might venture that either his value system was now different, or, if it was still the same, there was no way in which the old values could influence his decisions. No evidence exists to tell us which is true, yet my investigation of patients with brain damage similar to Phineas Gage's convinces me that neither explanation captures what really happens in those circumstances. Some part of the value system remains and can be utilized in abstract terms, but it is unconnected to real-life situations. When the Phineas Gages of this world need to operate in reality, the decision-making process is minimally influenced by old knowledge.

Another important aspect of Gage's story is the discrepancy between the degenerated character and the apparent intactness of the several instruments of mind—attention, perception, memory, language, intelligence. In this type of discrepancy, known in neuropsychology as *dissociation*, one or more performances within a general profile of operations are at odds with the rest. In Gage's case the impaired character was dissociated from the otherwise intact cognition and behavior. In other patients, with lesions elsewhere in the brain, language may be the impaired aspect, while character and all other cognitive aspects remain intact; language is then the "dissociated" ability. Subsequent study of patients similar to Gage has confirmed that his specific dissociation profile occurs consistently.

It must have been hard to believe that the character change would not resolve itself, and at first even Dr. Harlow resisted admitting that the change was permanent. This is understandable, since the most dramatic elements in Gage's story were his very survival, and then his survival without a defect that would more easily meet the eye: paralysis, for example, or a speech defect, or memory loss. Somehow, emphasizing Gage's newly developed social shortcomings smacked of ingratitude to both providence and medicine. By 1868, however, Dr. Harlow was ready to acknowledge the full extent of his patient's personality change.

Gage's survival was duly noted, but with the caution reserved for freakish phenomena. The significance of his behavioral changes was largely lost. There were good reasons for this neglect. Even in the small world of brain science at the time, two camps were beginning to form. One held that psychological functions such as language or memory could never be traced to a particular region of the brain. If one had to accept, reluctantly, that the brain did produce the mind, it did so as a whole and not as a collection of parts with special functions. The other camp held that, on the contrary, the brain did have specialized parts and those parts generated separate mind functions. The rift between the two camps was not merely indicative of the infancy of brain research; the argument endured for another century and, to a certain extent, is still with us today.

Whatever scientific debate Phineas Gage's story elicited, it focused on the issue of localizing language and movement in the brain. The debate never turned to the connection between impaired social conduct and frontal lobe damage. I am reminded here of a saying of Warren McCulloch's: "When I point, look where I point, not at my finger." (McCulloch, a legendary neurophysiologist and a pioneer in the field that would become computational neuroscience, was also a poet and a prophet. This saying was usually part of a prophecy.) Few looked to where Gage was unwittingly pointing. It is of course difficult to imagine anybody in Gage's day with the knowledge *and* the courage to look in the proper direction. It was acceptable that the brain sectors whose damage would have caused Gage's heart to stop pumping and his lungs to stop breathing had not been touched by the iron rod. It was also acceptable that the brain sectors which control wakefulness were far from the iron's course and were thus spared. It was even acceptable that the injury did not render Gage unconscious for a long period. (The event anticipated what is current knowledge from studies of head injuries: The style of the injury is a critical variable. A severe blow to the head, even if no bone is broken and no weapon penetrates the brain, can cause a major disruption of wakefulness for a long time; the forces unleashed by the blow disorganize brain function profoundly. A penetrating injury in which the forces are concentrated on a narrow and steady path, rather than dissipate and accelerate the brain against the skull, may cause dysfunction only where brain tissue is actually destroyed, and thus spare brain function elsewhere.) But to understand Gage's behavioral change would have meant believing that normal social conduct required a particular corresponding brain region, and this concept was far more unthinkable than its equivalent for movement, the senses, or even language.

Gage's case was used, in fact, by those who did not believe that mind functions could be linked to specific brain areas. They took a cursory view of the medical evidence and claimed that if such a wound as Gage's could fail to produce paralysis or speech impairments, then it was obvious that neither motor control nor language could be traced to the relatively small brain regions that neurologists had identified as motor and language centers. They argued—in complete error, as we shall see—that Gage's wound directly damaged those centers.[7]

The British physiologist David Ferrier was one of the few to take the trouble to analyze the findings with competence and wisdom.[8]

Ferrier's knowledge of other cases of brain lesion with behavioral changes, as well as his own pioneering experiments on electrical stimulation and ablation of the cerebral cortex in animals, had placed him in a unique position to appreciate Harlow's findings. He concluded that the wound spared motor and language "centers," that it did damage the part of the brain he himself had called the prefrontal cortex, and that such damage might be related to Gage's peculiar change in personality, to which Ferrier referred, picturesquely, as "mental degradation." The only supportive voices Harlow and Ferrier may have heard, in their very separate worlds, came from the followers of phrenology.

An Aside on Phrenology

What came to be known as phrenology began its days as "organology" and was founded by Franz Joseph Gall in the late 1700s. First in Europe, where it enjoyed a succès de scandale in the intellectual circles of Vienna, Weimar, and Paris, and then in America, where it was introduced by Gall's disciple and onetime friend Johann Caspar Spurzheim, phrenology sailed forth as a curious mixture of early psychology, early neuroscience, and practical philosophy. It had a remarkable influence in science and in the humanities, throughout most of the nineteenth century, although the influence was not widely acknowledged and the influenced took care to distance themselves from the movement.

Some of Gall's ideas are indeed quite astounding for the time. In no uncertain terms he stated that the brain was the organ of the spirit. With no less certitude he asserted that the brain was an aggregate of many organs, each having a specific psychological faculty. Not only did he part company with the favored dualist thinking, which separated biology from mind altogether, but he correctly intuited that there were many parts to this thing called brain, and that there was specialization in terms of the functions played by those parts.[9] The latter was a fabulous intuition since brain specialization is now a well-confirmed fact. Not surprisingly, however, he did not realize that the function of each separate brain part is not independent and that it is, rather, a contribution to the function of larger systems composed of those separate parts. But one can hardly fault Gall on this matter. It has taken the better part of two centuries for a "modern" view to take some hold. We can now say with confidence that there are no single "centers" for vision, or language, or for that matter, reason or social behavior. There are "systems" made up of several interconnected brain units; anatomically, but not functionally, those brain units are none other than the old "centers" of phrenologically inspired theory; and these systems are indeed dedicated to relatively separable operations that constitute the basis of mental functions. It is also true that the separate brain units, by virtue of where they are placed in a system, contribute different components to the system's operation and are thus not interchangeable. This is most important: What determines the contribution of a given brain unit to the operation of the system to which it belongs is not just the structure of the unit but also its *place* in the system.

The whereabouts of a unit is of paramount importance. This is why throughout this book I will talk so much about neuroanatomy, or brain anatomy, identify different brain regions, and even ask you to suffer the repeated mention of their names and the names of other regions with which they are interconnected. On numerous occasions I will refer to the presumed function of given brain regions, but such references should be taken in the context of the systems to which those regions belong. I am not falling into the phrenological trap. To put it simply: The mind results from the operation of each of the separate components, and from the concerted operation of the multiple systems constituted by those separate components.

While we must credit Gall with the concept of brain specialization, an impressive idea indeed given the scarce knowledge of his time, we must blame him for the notion of brain "centers" that he inspired. Brain centers became indelibly associated with "mental functions" in the work of nineteenth-century neurologists and physiologists. We also must be critical of various wild claims of phrenology, for instance, the idea that each separate brain "organ" generated mental faculties that were proportional to the size of the organ, or that all organs and faculties were innate. The notion of size as an index of the "power" or "energy" of a given mental faculty is amusingly wrong, although some contemporary neuroscientists have not shied away from using precisely the same notion in their work. The extension of this claim, the one that most undermined phrenology—and that many people think of when they hear the word—was that the organs could be identified from the outside by telltale bumps in the skull. As for the idea that organs and faculties are innate, you can see its influence throughout the nineteenth century, in literature as well as elsewhere; the magnitude of its error will be discussed in chapter 5.

The connection between phrenology and Phineas Gage's story deserves special mention. In his search for evidence about Gage, the psychologist M. B. MacMillan[10] uncovered a lead about one Nelson Sizer, a figure in phrenological circles of the 1800s who lectured in New England and who visited Vermont in the early 1840s, before Gage's accident. Sizer met John Harlow in 1842. In his otherwise rather boring book,[11] Sizer writes that "Dr. Harlow was then a young physician and assisted as a member of the committee at our lectures on phrenology in 1842." There were several followers of phrenology at medical schools in the eastern United States then, and Harlow was well acquainted with their ideas. He may have heard them speak in Philadelphia, a phrenology haven, or in New Haven or Boston, where Spurzheim had come in 1832, shortly after Gall's death, to be hailed as scientific leader and social sensation. New England wined and dined the hapless Spurzheim to the grave. His premature death came in a matter of weeks, although gratitude followed: the very night of the funeral, the Boston Phrenological Society was founded.

Whether or not Harlow ever heard Spurzheim, it is tantalizing to learn that he had at least one phrenology lesson directly from Nelson Sizer while the latter visited Cavendish (where he stayed—where else—at Mr. Adams's hotel). This influence may well explain Harlow's bold conclusion that Gage's behavioral transformation was due to a specific brain lesion and not to a general reaction to the accident. Intriguingly, Harlow does not rely on phrenology to support his interpretations.

Sizer did come back to Cavendish (and stayed again at Mr. Adams's hotel—in Gage's recovery room, naturally), and he was well aware of Gage's story. When Sizer wrote his book on phrenology in 1882, Phineas Gage was mentioned: "We perused [Harlow's] history of the case in 1848 with intense and affectionate interest, and also do not forget that the poor patient was quartered at the same hotel and in the same room."[11] Sizer's conclusion was that the iron bar had passed "in the neighborhood of Benevolence and the front part of Veneration." Benevolence and Veneration? Now, Benevolence and Veneration were not sisters in some Carmelite convent. They were phrenological "centers," brain "organs." Benevolence and Veneration gave people proper behavior, kindness and respect for other persons. Armed with this knowledge, you can understand Sizer's final view of Gage: "His organ of Veneration seemed to have been injured, and the profanity was the probable result." How true!

A LANDMARK BY HINDSIGHT

There is no question that Gage's personality change was caused by a circumscribed brain lesion in a specific site. But that explanation would not be apparent until two decades after the accident, and it became vaguely acceptable only in this century. For a long time, most everybody, John Harlow included, believed that "the portion of the brain traversed, was, for several reasons, the best fitted of any part of the cerebral substance to sustain the injury"[12]: in other words, a part of the brain that did nothing much and was thus expendable. But nothing could be further from the truth, as Harlow himself realized. He wrote in 1868 that Gage's mental recovery "was only partial, his intellectual faculties being decidedly impaired, but not totally lost; nothing like dementia, but they were enfeebled in their manifestations, his mental operations being perfect in kind, but not in degree or quantity." The unintentional message in Gage's case was that observing social convention, behaving ethically, and making decisions advantageous to one's survival and progress require knowledge of rules and strategies *and* the integrity of specific brain systems. The problem with this message was that it lacked the evidence required

to make it understandable and definitive. Instead the message became a mystery and came down to us as the "enigma" of frontal lobe
function. Gage posed more questions than he gave answers.

To begin with, all we knew about Gage's brain lesion was that it
was probably in the frontal lobe. That is a bit like saying that Chicago
is probably in the United States—accurate but not very specific or
helpful. Granted that the damage was likely to involve the frontal
lobe, where exactly was it within that region? The left lobe? The
right? Both? Somewhere else too? As you will see in the next chapter,
new imaging technologies have helped us come up with the answer
to this puzzle.

Then there was the nature of Gage's character defect. How did the
abnormality develop? The primary cause, sure enough, was a hole in
the head, but that just tells why the defect arose, not how. Might a
hole anywhere in the frontal lobe have the same result? Whatever
the answer, by what plausible means can destruction of a brain
region change personality? If there are specific regions in the frontal
lobe, what are they made of, and how do they operate in an intact
brain? Are they some kind of "center" for social behavior? Are they
modules selected in evolution, filled with problem-solving algo-
rithms ready to tell us how to reason and make decisions? How do
these modules, if that is what they are, interact with the environment
during development to permit normal reasoning and decision making? Or are there in fact no such modules?

What were the mechanisms behind Gage's failure at decision
making? It might be that the knowledge required to reason through a
problem was destroyed or rendered inaccessible, so that he no longer
could decide appropriately. It is possible also that the requisite
knowledge remained intact and accessible but the strategies for
reasoning were compromised. If this was the case, which reasoning
steps were missing? More to the point, which steps are there for
those who are allegedly normal? And if we are fortunate enough to
glean the nature of some of these steps, what are their neural
underpinnings?

Intriguing as all these questions are, they may not be as important as those which surround Gage's status as a human being. May he be described as having free will? Did he have a sense of right and wrong, or was he the victim of his new brain design, such that his decisions were imposed upon him and inevitable? Was he responsible for his acts? If we rule that he was not, does this tell us something about responsibility in more general terms? There are many Gages around us, people whose fall from social grace is disturbingly similar. Some have brain damage consequent to brain tumors, or head injury, or other neurological disease. Yet some have had no overt neurological disease and they still behave like Gage, for reasons having to do with their brains or with the society into which they were born. We need to understand the nature of these human beings whose actions can be destructive to themselves and to others, if we are to solve humanely the problems they pose. Neither incarceration nor the death penalty—among the responses that society currently offers for those individuals—contribute to our understanding or solve the problem. In fact, we should take the question further and inquire about our own responsibility when we "normal" individuals slip into the irrationality that marked Phineas Gage's great fall.

Gage lost something uniquely human, the ability to plan his future as a social being. How aware was he of this loss? Might he be described as self-conscious in the same sense that you and I are? Is it fair to say that his soul was diminished, or that he had lost his soul? And if so, what would Descartes have thought had he known about Gage and had he had the knowledge of neurobiology we now have? Would he have inquired about Gage's pineal gland?

Two

Gage's Brain
Revealed

THE PROBLEM

AT ABOUT THE time of the Phineas Gage affair, the neurologists Paul Broca in France and Carl Wernicke in Germany captured the attention of the medical world with their studies of neurological patients with brain lesions. Independently, Broca and Wernicke each proposed that damage to a well-circumscribed area in the brain was the cause of newly acquired language disorders in these patients.[1] The impairment in language became known technically as aphasia. The lesions, Broca and Wernicke thought, were thus revealing the neural underpinnings of two different aspects of language processing in normals. Their proposals were controversial and there was no rush to endorse them but the world did listen. With some reluctance and with much amendment they gradually became accepted. Harlow's work on Gage, however, or David Ferrier's comments, for that matter, never received the same attention, and never fired the imagination of their colleagues in the same way.

There were several reasons why. Even if a philosophical bent allowed one to think of the brain as the basis for the mind, it was difficult to accept the view that something as close to the human soul as ethical judgment, or as culture-bound as social conduct, might depend significantly on a specific region of the brain. Then there was the fact that Harlow was an amateur compared with Professors Broca and Wernicke, and could not marshal the convincing evidence required to make his case. Nowhere was this more obvious than in the failure to provide a precise location for the brain damage. Broca could state with certainty where in the brain the damage was that had caused language impairment, or aphasia, in his patients. He had studied their brains at the autopsy table. Likewise Wernicke, who had seen at postmortem that a back portion of the left temporal lobe was partially destroyed in patients exhibiting a language impairment—and noted that the aspect of language faculties affected was other than that identified by Broca. Harlow had not been able to make any such observation. Not only did he have to venture a relationship between Gage's brain damage and his behavioral impairment, but he had to conjecture where the damage was in the first place. He could not prove to anybody's satisfaction that he was right about anything.

Harlow's predicament was made worse by Broca's recently published findings. Broca had shown that lesions in the left frontal lobe, third frontal gyrus, caused language impairment in his patients. The entry and exit of the iron suggested that the damage to Gage's brain

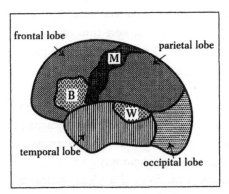

Figure 2-1. B = Broca area; M = motor area; W = Wernicke area. The four lobes are identified in the illustration. Harlow's critics claimed that Gage's lesion involved Broca's area, or the motor area, or even both, and used this claim to attack the idea that there was functional specialization in the human brain.

frontal lobe

parietal lobe

M

B

W

temporal lobe

occipital lobe

might be in the left frontal lobe. Yet Gage had no language impairment, while Broca's patients had no character defect. How could
there be such different results? With the scarce knowledge of functional neuroanatomy of the time, some people thought the lesions
were in approximately the same place, and that the different results
merely revealed the folly of those who wanted to find functional
specializations in the brain.

When Gage died in 1861, no autopsy was performed. Harlow
himself did not learn of Gage's death until about five years later. The
Civil War had been raging in the intervening years and news of this
sort did not travel fast. Harlow must have been saddened by Gage's
death and crushed at the lost opportunity of studying Gage's brain.
So crushed, in fact, that he proceeded to write Gage's sister with a
bizarre request. He petitioned her to have the body exhumed so that
the skull could be recovered and kept as a record of the case.

Phineas Gage was once again the involuntary protagonist of a grim
scene. His sister and her husband, D. D. Shattuck, along with a Dr.
Coon (then the mayor of San Francisco) and the family physician,
looked on as a mortician opened Gage's coffin and removed his skull.
The tamping iron, which had been placed alongside Gage's body, was
also retrieved, and sent with the skull to Dr. Harlow back East. Skull
and iron have been companions at the Warren Medical Museum of
the Harvard Medical School in Boston ever since.

For Harlow, being able to exhibit skull and iron was the closest he
could come to establishing that his case was not an invention, that a
man with such a wound had indeed existed. For Hanna Damasio,
some hundred twenty years later, Gage's skull was the springboard
for a piece of detective work that completed Harlow's unfinished
business and serves as a bridge between Gage and modern research
on frontal lobe function.

She began by considering the general trajectory of the iron, a
curious exercise in itself. Entering from the left cheek upward into
the skull, the iron broke through the back of the left orbital cavity
(the eye socket) located immediately above. Continuing upward it
must have penetrated the front part of the brain close to the midline,

although it was difficult to say where exactly. Since it seems to have been angled to the right it may have hit the left side first, then some of the right as it traveled upward. The initial site of brain damage probably was the orbital frontal region, directly above the orbital cavities. In its travel, the iron would have destroyed some of the inner surface of the left frontal lobe and perhaps of the right frontal lobe. Finally, as it exited, the iron would have damaged some part of the dorsal, or back, region of the frontal lobe, on the left side for sure and perhaps also on the right.

The uncertainties of this conjecture were obvious. There was a range of potential trajectories the iron might have taken through a "standard," idealized brain, and no way of knowing whether or how that brain resembled Gage's. The problem was made worse because although neuroanatomy jealously preserves topological relationships among its components, there are considerable degrees of individual topographic variation that make each of our brains far more different than cars of the same make. This point is best illustrated with the paradoxical sameness and difference of human faces: Faces have an invariant number of components and an invariant spatial arrangement (the topological relations of the components are the same in all human faces). Yet they are infinitely diverse and individually distinguishable because of small anatomical differences in size, contour, and position of those invariant parts and configuration (the precise topography changes from face to face). Individual brain variation, then, increased the likelihood that the above conjecture was erroneous.

Hanna Damasio proceeded to take advantage of modern neuroanatomy and state-of-the-art neuroimaging technology.[2] Specifically, she used a new technique she developed to reconstruct brain images of living humans in three dimensions. The technique, known as Brainvox,[3] relies on computer manipulation of raw data obtained from high-resolution magnetic resonance scans of the brain. In living normals or in neurological patients, it renders an image of the brain that is in no way different from the picture of that brain that you would be able to see at the autopsy table. It is an eerie, disquieting marvel. Think of what Prince Hamlet would have done, if he had been allowed to contemplate his own three pounds of brooding, indecisive brain, rather than just the empty skull the gravedigger handed him.

An Aside on the Anatomy of Nervous Systems

It may be useful here to outline the anatomy of the human nervous system. Why should any time be spent on this? In the previous chapter, when I discussed phrenology and the connection between brain structure and function, I mentioned the importance of neuroanatomy or brain anatomy. I emphasize it again because

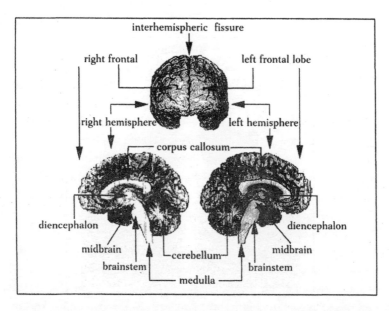

Figure 2-2. Human living brain reconstructed in three dimensions. The top center image shows the brain seen from the front. The corpus callosum is hidden underneath the interhemispheric fissure. The bottom images at the left and at the right show the two hemispheres of the same brain, separated at the middle as in a split-brain operation. The main anatomical structures are identified in the figure. The convoluted cover of the cerebral hemispheres is the cerebral cortex.

neuroanatomy is the fundamental discipline in neuroscience, from the level of microscopic single neurons (nerve cells) to that of the macroscopic systems spanning the entire brain. There can be no hope of understanding the many levels of brain function if we do not have a detailed knowledge of brain geography at multiple scales.

When we consider the nervous system in its entirety we can separate its central and peripheral divisions easily. The three-dimensional reconstruction in figure 2-2 represents the cerebrum, the main component of the central nervous system. In addition to the cerebrum, with its left and right cerebral hemispheres joined by the corpus callosum (a thick collection of nerve fibers connecting left and right hemispheres bidirectionally), the central nervous system includes the

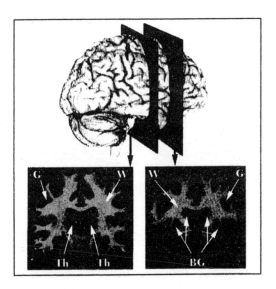

Figure 2-3. Two sections through a reconstructed living human brain obtained with magnetic resonance imaging (MRI) and the Brainvox technique. The planes of section are identified in the image at the top and center. The difference between gray (G) and white matter (W) is readily visible. Gray matter shows up in the cerebral cortex, the gray ribbon which makes up the entire contour of every hump and crevice in the section, and in deep nuclei such as the basal ganglia (BG) and the thalamus (Th).

diencephalon (a midline collection of nuclei, hidden under the hemi-spheres, which includes the thalamus and the hypothalamus), the midbrain, the brain stem, the cerebellum, and the spinal cord.

The central nervous system is "neurally" connected to almost every nook and cranny of the remainder of the body by nerves, the collec-tion of which constitute the peripheral nervous system. Nerves ferry impulses from brain to body and from body to brain. As will be discussed in chapter 5, however, brain and body are also intercon-nected chemically, by substances such as hormones and peptides, which are released in one and go to the other via the bloodstream.

When we section the central nervous system we can make out with-out difficulty the difference between its dark and pale sectors. (Figure 2-3). The dark sectors are known as the gray matter although their real color is usually brown rather than gray. The pale sectors are known as the white matter. The gray matter corresponds largely to collections of nerve cell bodies, while the white matter corresponds largely to axons, or nerve fibers, emanating from cell bodies in the gray matter.

The gray matter comes in two varieties. In one variety the neurons are layered as in a cake and form a *cortex*. Examples are the cerebral cortex which covers the cerebral hemispheres, and the cerebellar cortex which envelops the cerebellum. In the second variety of gray matter the neurons are not layered and are organized instead like

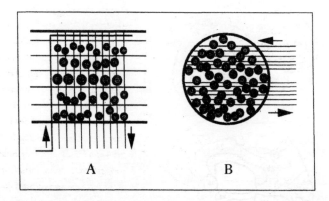

A   B

Figure 2-4. A = diagram of the cellular architecture of cerebral cortex with its charac-teristic layer structure; B = diagram of the cellular architecture of a nucleus.

cashew nuts inside a bowl. They form a *nucleus*. There are large nuclei, such as the caudate, putamen, and pallidum, quietly hidden in the depth of each hemisphere; or the amygdala, hidden inside each temporal lobe; there are large collections of smaller nuclei, such as those that form the thalamus; and small individual nuclei, such as the substantia nigra or the nucleus ceruleus, located in the brain stem.

The brain structure to which neuroscience has dedicated the most effort is the cerebral cortex. It can be visualized as a comprehensive mantle to the cerebrum, covering all its surfaces, including those located in the depth of crevices known as fissures and sulci which give the brain its characteristic folded appearance. (See Fig. 2-2.) The thickness of this multilayer blanket is about 3 millimeters, and the layers are parallel to one another and to the brain's surface. (See Fig. 2-4). All gray matter below the cortex (nuclei, large and small, and the cerebellar cortex) is known as subcortical. The evolutionarily modern part of the cerebral cortex is called the neocortex. Most of the evolutionarily older cortex is known as limbic cortex (see below). Throughout the book I will usually refer either to cerebral cortex (meaning neocortex), or to limbic cortex and its specific parts.

Figure 2-5 depicts a frequently used map of the cerebral cortex based on its varied cytoarchitectonic areas (regions of distinctive

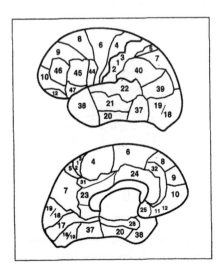

Figure 2-5. A map of the main brain areas identified by Brodmann in his studies of cellular architecture (cytoarchitectonics). This is neither a phrenology map nor a contemporary map of brain functions. It is simply a convenient anatomical reference. Some areas are too small to be depicted here, or they are hidden in the depth of sulci and fissures. The top image corresponds to the external aspect of the left hemisphere, and the bottom one to the internal aspect.

cellular architecture). It is known as Brodmann's map and its areas are designated by number.

One division of the central nervous system to which I will refer often is both cortical and subcortical and is known as the limbic system. (The term is something of a catchall for a number of evolutionarily old structures, and although many neuroscientists resist using it, it often comes in handy.) The main structures of the limbic system are the cingulate gyrus, in the cerebral cortex, and the amygdala and basal forebrain, two collections of nuclei.

The nervous (or neural) tissue is made up of nerve cells (neurons) supported by glial cells. Neurons are the cells essential for brain activity. There are billions of such neurons in our brains, organized in local circuits, which, in turn, constitute cortical regions (if they are arranged in layers) or nuclei (if they are aggregated in nonlayered collections). Finally, the cortical regions and nuclei are interconnected to form systems, and systems of systems, at progressively higher levels of complexity. In terms of scale, all neurons and local circuits are microscopic, while cortical regions, nuclei, and systems are macroscopic.

Neurons have three important components: a cell body; a main output fiber, the axon; and input fibers, or dendrites. (See Fig. 2-6)

Figure 2-6. Diagram of a neuron with its main components: cell body, dendrites, and portion of axon.

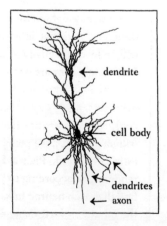
← dendrite

← cell body

← dendrites
← axon

Neurons are interconnected in circuits in which there are the equivalent of conducting wires (the neurons' axon fibers) and connectors (synapses, the points at which axons make contact with the dendrites of other neurons).

When neurons become active (a state known in neuroscience jargon as "firing"), an electric current is propagated away from the cell body and down the axon. This current is the action potential, and when it arrives at a synapse it triggers the release of chemicals known as neurotransmitters (glutamate is one such transmitter). In turn, neurotransmitters operate on receptors. In an excitatory neuron, the cooperative interaction of many other neurons whose synapses are adjacent and which may or not release their own transmitters, determines whether or not the next neuron will fire, that is, whether it will produce its own action potential, which will lead to its own neurotransmitter release, and so forth.

Synapses can be strong or weak. Synaptic strength decides whether or not, and how easily, impulses continue to travel into the next neuron. In general, in an excitatory neuron, a strong synapse facilitates impulse travel, while a weak synapse impedes or blocks it.[4]

A neuroanatomical issue I must mention to conclude this aside has to do with the nature of neuron connectivity. It is not uncommon to find scientists who despair of ever understanding the brain when they are confronted by the complexity of connections among neurons. Some prefer to hide behind the notion that everything connects with everything else and that mind and behavior probably emerge from that willy-nilly connectivity in ways that neuroanatomy will never reveal. Fortunately, they are wrong. Consider the following: On the average, every neuron forms about 1,000 synapses, although some can have as many as 5,000 or 6,000. This may seem a high number, but when we consider that there are more than 10 billion neurons and more than 10 *trillion* synapses, we realize that each neuron is nothing if not modestly connected. Pick a few neurons in the cortex or in nuclei, randomly or according to your anatomical preferences, and you will find that each neuron talks to a few others but never to most or all of the others. In fact, many neurons talk only to neurons that are not

Levels of Neural Architecture
Neurons
Local Circuits
Subcortical Nuclei
Cortical Regions
Systems
Systems of Systems

very far away, within relatively local circuits of cortical regions and nuclei, and others, although their axons sail forth for several millimeters, even centimeters, across the brain, will still make contact with only a relatively small number of other neurons. The main consequences of this arrangement are as follows: (1) whatever neurons do depends on the nearby assembly of neurons they belong to; (2) whatever systems do depends on how assemblies influence other assemblies in an architecture of interconnected assemblies; and (3) whatever each assembly contributes to the function of the system to which it belongs depends on its place in that system. In other words, the brain specialization mentioned in the aside on phrenology in chapter 1 is a consequence of the place occupied by assemblies of sparsely connected neurons within a large-scale system.

In short, then, the brain is a supersystem of systems. Each system is composed of an elaborate interconnection of small but macroscopic cortical regions and subcortical nuclei, which are made of microscopic local circuits, which are made of neurons, all of which are connected by synapses. (It is not uncommon to find the terms "circuit" and "network" used as synonyms of "system." To avoid confusion, it is important to specify whether a microscopic or macroscopic scale is intended. In this text, unless otherwise stated, systems are macroscopic and circuits are microscopic.)

THE SOLUTION

Since Phineas Gage was not around to be scanned, Hanna Damasio thought of an indirect approach to his brain.[5] She enlisted the help of Albert Galaburda, a neurologist at Harvard Medical School, who went to the Warren Medical Museum and carefully photographed Gage's skull from different angles, and measured the distances between the areas of bone damage and a variety of standard bone landmarks.

Analysis of these photographs combined with the descriptions of the wound helped narrow down the range of possible courses for the iron bar. The photographs also allowed Hanna Damasio and her neurologist colleague, Thomas Grabowski, to re-create Gage's skull in three-dimensional coordinates and to derive from them the most likely coordinates of the brain that best fitted such a skull. With the help of her collaborator Randall Frank, an engineer, Damasio then performed a simulation in a high-power computer work station. They re-created a three-dimensional iron rod with the precise dimensions of Gage's tamping iron, and "impaled" it on a brain whose shape and size were close to Gage's, along the now narrowed range of possible trajectories that the iron might have followed during the accident. The results are shown in Figures 2-7 and 2-8.

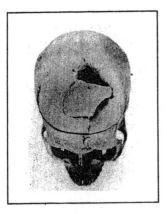

Figure 2-7. Photograph of Gage's skull obtained in 1992.

32 DESCARTES' ERROR

Figure 2-8. TOP
PANELS: A recon-
struction of Gage's
brain and skull with
the likely trajectory of
the iron rod marked in
dark gray.
BOTTOM PANELS:
A view of both left and
right hemispheres as
seen from the inside,
showing how the iron
damaged frontal lobe
structures on both
sides.

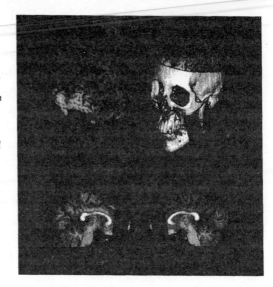

We can now confirm David Ferrier's claim that in spite of the
amount of brain lost, the iron did not touch the brain regions
necessary for motor function or language. (The intact areas of both
hemispheres included the motor and premotor cortices, as well as
the frontal operculum, on the left side known as Broca's area.) We
can state with confidence that the damage was more extensive on
the left than on the right hemisphere, and on the anterior than the
posterior sectors of the frontal region as a whole. The damage
compromised prefrontal cortices in the ventral and inner surfaces of
both hemispheres while preserving the lateral, or external, aspects
of the prefrontal cortices.

Part of a region which our recent investigations have highlighted
as critical for normal decision-making, the ventromedial prefrontal
region, was indeed damaged in Gage. (In neuroanatomical terminol-
ogy, the orbital region is known also as the *ventromedial* region of the
frontal lobe, and this is how I will refer to it throughout the book.
"Ventral" and "ventro-" come from *venter*, "belly" in Latin, and this
region is the underbelly of the frontal lobe, so to speak; "medial"

designates proximity to the midline or the inside surface of a structure.) The reconstruction revealed that regions thought to be vital for other aspects of neuropsychological function were not damaged in Gage. The cortices in the lateral aspect of the frontal lobe, for instance, whose damage disrupts the ability to control attention, perform calculations, and shift appropriately from stimulus to stimulus, were intact.

This modern research allowed certain conclusions. Hanna Damasio and her colleagues could say with some foundation that it was selective damage in the prefrontal cortices of Phineas Gage's brain that compromised his ability to plan for the future, to conduct himself according to the social rules he previously had learned, and to decide on the course of action that ultimately would be most advantageous to his survival. What was missing now was the knowledge of how Gage's mind might have worked when he behaved as dismally as he did. And for that we had to investigate the modern counterparts of Phineas Gage.

Three

A Modern
Phineas Gage

NOT LONG AFTER I began seeing patients whose behavior resembled Gage's and first became fascinated by the results of prefrontal damage—a full two decades ago—I was asked to see a patient with an especially pure version of the condition. The patient had undergone a radical change of personality, I was told, and the referring physicians had a special request: they wanted to know whether this change so at odds with previous behavior was a real disease. Elliot, as I will refer to the patient, was then in his thirties.[1] No longer capable of holding a job, he was living in the custody of a sibling and the pressing issue was that he was being denied payment of disability benefits. For all the world to see, Elliot was an intelligent, skilled, and able-bodied man who ought to come to his senses and return to work. Several professionals had declared that his mental faculties were intact—meaning that at the very best Elliot was lazy, and at the worst a malingerer.

I saw Elliot at once, and he struck me as pleasant and intriguing, thoroughly charming but emotionally contained. He had a respectful, diplomatic composure, belied by an ironic smile implying superior wisdom and a faint condescension with the follies of the world.

He was cool, detached, unperturbed even by potentially embarrass-
ing discussion of personal events. He reminded me somewhat of
Addison DeWitt, the character played by George Sanders in *All
About Eve*.

Not only was Elliot coherent and smart, but clearly he knew what
was occurring in the world around him. Dates, names, details in the
news were all at his fingertips. He discussed political affairs with
the humor they often deserve and seemed to grasp the situation of
the economy. His knowledge of the business realm he had worked in
remained strong. I had been told his skills were unchanged, and that
appeared plausible. He had a flawless memory for his life story,
including the most recent, strange events. And the strangest things
had indeed been happening.

Elliot had been a good husband and father, had a job with a
business firm, and had been a role model for younger siblings and
colleagues. He had attained an enviable personal, professional,
and social status. But his life began to unravel. He developed severe
headaches, and soon it was hard for him to concentrate. As his
condition worsened, he seemed to lose his sense of responsibility,
and his work had to be completed or corrected by others. His family
physician suspected that Elliot might have a brain tumor. Regrettably, the suspicion proved correct.

The tumor was large and growing fast. By the time it was diagnosed it had attained the size of a small orange. It was a meningioma,
so-called because it arises out of the membranes covering the brain's
surface, which are called meninges. I later learned that Elliot's
tumor had begun growing in the midline area, just above the nasal
cavities, above the plane formed by the roof of the eye sockets. As the
tumor grew bigger, it compressed both frontal lobes upward, from
below.

Meningiomas are generally benign, as far as the tumor tissue itself
is concerned, but if they are not removed surgically they can be just
as fatal as the tumors we call malignant. As they keep compressing
brain tissue in their growth, they eventually kill it. Surgery was
necessary if Elliot was to survive.

An excellent medical team performed the surgery, and the tumor was removed. As is usual in such cases, frontal lobe tissue that had been damaged by the tumor had to be removed too. The surgery was a success in every respect, and insofar as such tumors tend not to grow again, the outlook was excellent. What was to prove less felicitous was the turn in Elliot's personality. The changes, which began during his physical recovery, astonished family and friends. To be sure, Elliot's smarts and his ability to move about and use language were unscathed. In many ways, however, Elliot was no longer Elliot.

Consider the beginning of his day: He needed prompting to get started in the morning and prepare to go to work. Once at work he was unable to manage his time properly; he could not be trusted with a schedule. When the job called for interrupting an activity and turning to another, he might persist nonetheless, seemingly losing sight of his main goal. Or he might interrupt the activity he had engaged, to turn to something he found more captivating at that particular moment. Imagine a task involving reading and classifying documents of a given client. Elliot would read and fully understand the significance of the material, and he certainly knew how to sort out the documents according to the similarity or disparity of their content. The problem was that he was likely, all of a sudden, to turn from the sorting task he had initiated to reading one of those papers, carefully and intelligently, and to spend an entire day doing so. Or he might spend a whole afternoon deliberating on which principle of categorization should be applied: Should it be date, size of document, pertinence to the case, or another? The flow of work was stopped. One might say that the particular step of the task at which Elliot balked was actually being carried out *too well*, and at the expense of the overall purpose. One might say that Elliot had become irrational concerning the larger frame of behavior, which pertained to his main priority, while within the smaller frames of behavior, which pertained to subsidiary tasks, his actions were unnecessarily detailed.

His knowledge base seemed to survive, and he could perform many separate actions as well as before. But he could not be counted on to perform an appropriate action when it was expected. Under-standably, after repeated advice and admonitions from colleagues and superiors went unheeded, Elliot's job was terminated. Other jobs—and other dismissals—were to follow. Elliot's life was now beating to a different drum.

No longer tied to regular employment, Elliot charged ahead with new pastimes and business ventures. He developed a collecting habit—not a bad thing in itself, but less than practical when the collected objects were junk. The new businesses ranged from home-building to investment management. In one enterprise, he teamed up with a disreputable character. Several warnings from friends were of no avail, and the scheme ended in bankruptcy. All of his savings had been invested in the ill-fated enterprise and all were lost. It was puzzling to see a man with Elliot's background make such flawed business and financial decisions.

His wife, children, and friends could not understand why a knowl-edgeable person who was properly forewarned could act so foolishly, and some among them could not cope with this state of affairs. There was a first divorce. Then a brief marriage to a woman of whom neither family nor friends approved. Then another divorce. Then more drifting, without a source of income, and as a final blow to those who still cared and were watching in the sidelines, the denial of social security disability payments.

Elliot's benefits were restored. I explained that his failures were indeed caused by a neurological condition. True, he was still phys-ically capable and most of his mental capacities were intact. But his ability to reach decisions was impaired, as was his ability to make an effective plan for the hours ahead of him, let alone to plan for the months and years of his future. These changes were in no way comparable to the slips of judgment that visit all of us from time to time. Normal and intelligent individuals of comparable education make mistakes and poor decisions, but not with such systematically dire consequences. The changes in Elliot had a larger magnitude and were a sign of disease. Nor were these changes consequent to a former weakness of character, and they certainly were not controlled willfully by the patient; their root cause, quite simply, was damage to a particular sector of the brain. Furthermore, the changes had a chronic character. Elliot's condition was not transient. It was there to stay.

The tragedy of this otherwise healthy and intelligent man was that he was neither stupid nor ignorant, and yet he acted often as if he were. The machinery for his decision making was so flawed that he could no longer be an effective social being. In spite of being confronted with the disastrous results of his decisions, he did not learn from his mistakes. He seemed beyond redemption, like the repeat offender who professes sincere repentance but commits another offense shortly thereafter. It is appropriate to say that his free will had been compromised and to venture, in answer to the question I had posed concerning Gage, that Gage's free will had been compromised too.

In some respects Elliot was a new Phineas Gage, fallen from social grace, unable to reason and decide in ways conducive to the maintenance and betterment of himself and his family, no longer capable of succeeding as an independent human being. And like Gage he had even developed a collecting habit. In other respects, however, Elliot was different. He was less intense than Gage appears to have been, and he never used profanity. Whether the differences correspond to slightly different locations of their respective lesions, or to differences in sociocultural background, premorbid personality, or age, is an empirical question for which I do not yet have the answer.

Even before studying Elliot's brain with modern imaging techniques, I knew that the damage involved the frontal lobe region; his neuropsychological profile indicated this region alone. As we will see in chapter 4, damage in other sites (in the right-side somatosensory cortex, for instance) can compromise decision making, but in such cases there are other accompanying defects (major paralysis, disturbance of the processing of sensation).

The computerized tomography and magnetic resonance studies performed on Elliot revealed that both the right and the left frontal lobes had suffered, and that the damage was far greater on the right than on the left. In fact, the external surface of the left frontal lobe was intact, and all damage on the left side was within the orbital and medial sectors. On the right side, these sectors were similarly damaged, but in addition the core of the lobe (the white matter under the cerebral cortex) was destroyed. As a result of the destruction, a large component of the right frontal cortices was not functionally viable.

On both sides, the parts of the frontal lobe concerned with controlling movement (the motor and premotor regions) were not damaged. This was not surprising, since Elliot's movements were entirely normal. Also, as expected, the frontal language-related cortices (Broca's area and its surroundings) were intact. The region just behind the base of the frontal lobe, the basal forebrain, was likewise intact. That region is one of several necessary for learning and memory. Had it been damaged, Elliot's memory would have been impaired.

Was there evidence of any other damage in Elliot's brain? The answer is a definite no. The temporal, occipital, and parietal regions were intact in both left and right hemispheres. The same was true of the large gray-matter nuclei beneath the cortex, the basal ganglia and the thalamus. The damage was thus confined to prefrontal cortices. Just as in Gage, the ventromedial sector of those cortices had taken a disproportionate brunt of damage. The damage to Elliot's brain, though, was more extensive on the right than the left.

Little brain was destroyed, one might think; much was left intact. Yet amount of damage is often not the point as far as the consequences of brain damage are concerned. The brain is not one big lump of neurons doing the same thing wherever they are. The structures destroyed in both Gage and Elliot happened to be those necessary for reasoning to culminate in decision making.

A NEW MIND

I remember being impressed by Elliot's intellectual soundness, but I remember also thinking that other patients with frontal lobe damage *seemed* sound when they had in fact subtle changes in intellect, detectable only by special neuropsychological tests. Their altered behavior often had been attributed to defects in memory or attention. Elliot would disabuse me of that notion.

He had been evaluated previously at another institution where the opinion had been that there was no evidence of "organic brain syndrome." In other words, he showed no sign of impairment when he was given standard intelligence tests. His intelligence quotient (the so-called IQ) was in the superior range, and his standing on the Wechsler Adult Intelligence Scale indicated no abnormality. His problems were found not to result from "organic disease" or "neurological dysfunction"—in other words, brain disease—but instead to reflect "emotional" and "psychological" adjustment problems—in other words, mental trouble—and would be thus amenable to psychotherapy. Only after a series of therapy sessions proved unsuccessful was Elliot referred to our unit. (The distinction between diseases of "brain" and "mind," between "neurological" problems and "psychological" or "psychiatric" ones, is an unfortunate cultural inheritance that permeates society and medicine. It reflects a basic ignorance of the relation between brain and mind. Diseases of the brain are seen as tragedies visited on people who cannot be blamed for their condition, while diseases of the mind, especially those that affect conduct and emotion, are seen as social inconveniences for which sufferers have much to answer. Individuals are to be blamed for their character flaws, defective emotional modulation, and so on; lack of willpower is supposed to be the primary problem.)

The reader may well ask whether the previous medical evaluation was in error. Is it conceivable that somebody as impaired as Elliot would perform well on psychological tests? In fact it is: patients with marked abnormalities of social behavior can perform normally on many and even most intelligence tests, and clinicians and investigators have struggled for decades with this frustrating reality. There may be brain disease, but laboratory tests fail to measure significant impairments. The problem here lies with the tests, not with the patients. The tests simply do not address properly the particular functions that are compromised and thus fail to measure any decline. Knowing of Elliot's condition and his lesion, I predicted that he would be found normal on most psychological tests but abnormal on a small number of tests which are sensitive to malfunction in frontal cortices. As you will see, Elliot would surprise me.

The standardized psychological and neuropsychological tests revealed a superior intellect.[2] On every subtest of the Wechsler Adult Intelligence Scale, Elliot showed abilities that were either superior or average. His immediate memory for digits was superior, as were his short-term verbal memory and visual memory for geometric designs. His delayed recall of Rey's word list and complex figures were in the normal range. His performance on the Multilingual Aphasia Examination, a battery of tests which assesses various aspects of language comprehension and production, was normal. His visual perception and construction skills were normal on Benton's standardized tests of facial discrimination, judgment of line orientation, tests of geographic orientation, and two- and three-dimensional block construction. The copy of the Rey-Osterrieth complex figure was also normal.

Elliot performed normally on memory tests employing interference procedures. One test involved the recall of consonant trigrams after three-, nine-, and eighteen-second delays, with the distraction of counting backward; another, the recall of items after a fifteen-second delay spent in calculations. Most patients with frontal lobe damage test abnormally; Elliot performed well in both tasks, with 100 and 95 percent accuracy, respectively.

In short, perceptual ability, past memory, short-term memory, new learning, language, and the ability to do arithmetic were intact. Attention, the ability to focus on a particular mental content to the exclusion of others, was also intact; and so was working memory, which is the ability to hold information in mind over a period of many seconds and to operate on it mentally. Working memory is usually tested in the domains of words or numbers, objects or their features. For example, after being told of a telephone number, the subject will be asked to repeat it immediately afterward in backward direction, skipping the odd digits.

My prediction that Elliot would fail on tests known to detect
frontal lobe dysfunction was not correct. He turned out to be so
intact intellectually that even the special tests were a breeze for him.
The task to be given was the Wisconsin Card Sorting Test, the
workhorse of the small group of so-called frontal lobe tests, which
involves sorting through a long series of cards whose face image can
be categorized according to color (e.g., red or green), shape (stars,
circles, squares), and number (one, two, or three elements). When
the examiner shifts the criterion according to which the subject is
sorting, the subject must realize the change quickly and switch to the
new criterion. In the 1960s the psychologist Brenda Milner showed
that patients with damage to prefrontal cortices often are impaired
in this task, and this finding has been confirmed repeatedly by other
investigators.[3] Patients tend to stick to one criterion rather than shift
gears appropriately. Elliot achieved six categories in seventy sorts—
something that most patients with frontal lobe damage cannot do.
He sailed through the task, seemingly no different from unimpaired
people. Through the years he has maintained this type of performance on the Wisconsin test and on comparable tasks. Implicit in
Elliot's normal performance in this test are the ability to attend and
operate on a working memory, as well as an essential logical competence and the ability to change mental set.

The ability to make estimates on the basis of incomplete knowledge is another index of superior intellectual function that is often
compromised in patients with frontal lobe damage. Two researchers,
Tim Shallice and M. Evans, have devised a task to assess this ability
consisting of questions for which you will not have a precise answer
(unless, perhaps, you are a collector of trivia), and which can be
answered only by conjuring up a variety of unconnected facts, and
operating on them with logical competence so as to arrive at a valid
inference.[4] Imagine being asked, for example, how many giraffes
there are in New York City, or how many elephants in the state of
Iowa. You must consider that neither species is indigenous to North
America, and that zoos and wild life parks are thus the only place
where they can be found; you must also consider the overall map of

New York City or the state of Iowa, and plot how many such facilities are likely to exist in each space; and from another bank of your knowledge you may estimate the probable number of giraffes and elephants in *each* such facility; and eventually add it all up and come up with a number.) I hope you answer with a reasonable ballpark figure; but I would be surprised—and worried—if you know the exact number). In essence you have to generate an acceptable estimate based on bits and pieces of unrelated knowledge; and you must have normal logical competence, normal attention, and normal working memory. It is of interest to know, then, that the often unreasonable Elliot produced cognitive estimates in the normal range.

By then Elliot had passed through most of the hoops set up for him. He had not taken a personality test yet, and this would be it, I thought. What was the chance that he would fare well in the prime personality test, the Minnesota Multiphasic Personality Inventory,[5] also known as MMPI. As you may have guessed by now, Elliot was normal in that one too. He generated a valid profile; his performance was genuine.

After all these tests, Elliot emerged as a man with a normal intellect who was unable to decide properly, especially when the decision involved personal or social matters. Could it be that reasoning and decision making in the personal and social domain were different from reasoning and thinking in domains concerning objects, space, numbers, and words? Might they depend on different neural systems and processes? I had to accept the fact that despite the major changes that had followed his brain damage, nothing much could be measured in the laboratory with the traditional neuropsychological instruments. Other patients had shown this sort of dissociation, but none so devastatingly, as far as we investigators were concerned. If we were to measure any impairment, we had to develop new approaches. And if we wanted to explain Elliot's behavior defects satisfactorily, we should desist from the traditional accounts; Elliot's impeccable performances meant that the usual suspects could not be blamed.

RESPONDING TO THE CHALLENGE

Few things can be as salutary, once you find an intellectual hurdle, as giving yourself a vacation from the problem. So I took some time off from the problem of Elliot, and when I returned, I found that my perspective on the case had begun to change. I realized I had been overly concerned with the state of Elliot's intelligence and the instruments of his rationality, and had not paid much attention to his emotions, for various reasons. At first glance, there was nothing out of the ordinary about Elliot's emotions. He was, as I said earlier, an emotionally contained sort, but many illustrious and socially exemplary people have been emotionally contained. He certainly was not overemotional; he did not laugh or cry inappropriately, and he seemed neither sad nor joyful. He was not facetious, just quietly humorous (his wit was far more engaging and socially acceptable than that of some people I know). On a more probing analysis, however, something was missing, and I had overlooked much of the prime evidence for this: Elliot was able to recount the tragedy of his life with a detachment that was out of step with the magnitude of the events. He was always controlled, always describing scenes as a dispassionate, uninvolved spectator. Nowhere was there a sense of his own suffering, even though he was the protagonist. Mind you, restraint of this sort is often most welcome, from the point of view of a physician-listener, since it does reduce one's emotional expense. But as I talked to Elliot again for hours on end, it became clear that the magnitude of his distance was unusual. Elliot was exerting no restraint whatsoever on his affect. He was calm. He was relaxed. His narratives flowed effortlessly. He was not inhibiting the expression of internal emotional resonance or hushing inner turmoil. He simply did not have any turmoil to hush. This was not a culturally acquired stiff upper lip. In some curious, unwittingly protective way, he was not pained by his tragedy. I found myself suffering more when listening to Elliot's stories than Elliot himself seemed to be suffering. In fact, I felt that I suffered more than he did just by *thinking* of those stories.

Bit by bit the picture of this disaffection came together, partly from my observations, partly from the patient's own account, partly from the testimony of his relatives. Elliot was far more mellow in his emotional display now than he had been before his illness. He seemed to approach life on the same neutral note. I never saw a tinge of emotion in my many hours of conversation with him: no sadness, no impatience, no frustration with my incessant and repetitious questioning. I learned that his behavior was the same in his own daily environment. He tended not to display anger, and on the rare occasions when he did, the outburst was swift; in no time he would be his usual new self, calm and without grudges.

Later, and quite spontaneously, I would obtain directly from him the evidence I needed. My colleague Daniel Tranel had been conducting a psychophysiological experiment in which he showed subjects emotionally charged visual stimuli—for instance, pictures of buildings collapsing in earthquakes, houses burning, people injured in gory accidents or about to drown in floods. As we debriefed Elliot from one of many sessions of viewing these images, he told me without equivocation that his own feelings had changed from before his illness. He could sense how topics that once had evoked a strong emotion no longer caused any reaction, positive or negative.

This was astounding. Try to imagine it. Try to imagine not feeling pleasure when you contemplate a painting you love or hear a favorite piece of music. Try to imagine yourself forever robbed of that possibility and yet aware of the intellectual contents of the visual or musical stimulus, and also aware that once it did give you pleasure. We might summarize Elliot's predicament as *to know but not to feel*.

I became intrigued with the possibility that reduced emotion and feeling might play a role in Elliot's decision-making failures. But further studies, of Elliot and other patients, were necessary to support this idea. I needed, first of all, to exclude beyond the shadow of a doubt that I had not missed detecting any primary intellectual difficulty, one that might explain Elliot's problems independently of any other defect.

REASONING AND DECIDING

The continued exclusion of subtle intellectual defects took many paths. It was important to establish whether Elliot still knew the rules and principles of behavior that he neglected to use day after day. In other words, had he lost knowledge concerning social behavior, so that even with his normal reasoning mechanisms he would not be able to solve a problem? Or was he still in possession of the knowledge but no longer able to conjure it up and manipulate it? Or was he able to gain access to the knowledge but unable to operate on it and make a choice?

I was helped in this investigation by my then student Paul Eslinger. We began by presenting Elliot with a series of problems, centered on ethical dilemmas and financial questions. Say he needed cash, for example; would he steal if given the opportunity and the virtual guarantee that he would not be discovered? Or: If he knew the performance of company X's stock over the past month, would he sell any stock he owned or buy more of it? Elliot responded no differently from how any of us in the laboratory would have. His ethical judgments followed principles we all shared. He was aware of how social conventions applied to the problems. His financial decisions sounded reasonable. There was nothing especially sophisticated about the problems we set, but it was remarkable to discover, nonetheless, that Elliot did not perform abnormally. His real-life performance, after all, was a catalogue of violations in the domains covered by the problems. This dissociation between real-life failure and laboratory normalcy presented yet another challenge.

My colleague Jeffrey Saver would later respond to this challenge by studying Elliot's behavior in a series of controlled laboratory tasks having to do with social convention and moral value. Let me describe the tasks.

The first concerned the generation of options for action. This instrument was designed to measure the ability to devise alternative solutions to hypothetical social problems. Four social situations (predicaments, in fact) are presented verbally in the test, and the subject is asked to produce different verbal-response options to each (which he is supposed to describe verbally). In one situation, the protagonist breaks a spouse's flower pot; the subject is asked to come up with actions the protagonist might take to prevent the spouse from becoming angry. A standardized set of questions such as "What else can he do?" is employed to elicit alternative solutions. The number of relevant and discrete solutions conceptualized by the subject are scored before and after prompting. Elliot exhibited no deficit in performance relative to that of a control group in number of relevant solutions generated prior to prompting, total number of relevant solutions, or relevance score.

The second task concerned awareness of consequences. This measure was constructed to sample a subject's spontaneous inclination to consider the consequences of actions. The subject is presented with four hypothetical situations in which there arises a temptation to transgress ordinary social convention. In one segment, the protagonist cashes a check at a bank and is given too much money by the teller. The subject is asked to describe how the scenario might evolve, and indicate the protagonist's thoughts prior to an action and any subsequent thoughts or events. The subject's score reflects the frequency with which his or her replies include a consideration of the consequences of choosing a particular option. On this task Elliot's performance was even superior to that of the control group.

The third task, the Means-Ends Problem-Solving Procedure, concerned the ability to conceptualize efficacious means of achieving a social goal. The subject is given ten different scenarios and is to conceive appropriate and effective measures to reach a specified goal in order to satisfy a social need—for instance, forming a friendship, maintaining a romantic relationship, or resolving an occupational difficulty. The subject might be told about someone who moves to a new neighborhood, and develops many good friends and feels at home there. The subject then is asked to elaborate a story describing the events that led to this successful outcome. The score is the number of effective acts leading to the outcome. Elliot performed impeccably.

The fourth task concerned the ability to predict the social consequences of events. In each of the thirty test items, the subject views a cartoon panel showing an interpersonal situation, and is asked to choose from among three other panels the one that depicts the most likely outcome of the initial panel. Scoring reflects the number of correct choices. Elliot was no different from normal control subjects.

The fifth and final task, the Standard Issue Moral Judgment Interview (a modified version of the Heinz dilemma as designed by L. Kohlberg and colleagues),[6] concerned the developmental stage of moral reasoning. Presented with a social situation that poses a conflict between two moral imperatives, the subject is asked to indicate a solution to the dilemma and to provide a detailed ethical justification for that solution. In one such situation, for instance, the subject must decide, and explain, whether or not a character should steal a drug to prevent his wife from dying. Scoring employs explicit staging criteria to assign each interview judgment to a specific level of moral development.

The Standard Issue Moral Judgment Interview score ranks a subject in one of five successively more complex stages of moral reasoning. These modes of moral reasoning include preconventional levels (stage 1, obedience and punishment orientation; stage 2, instrumental purpose and exchange); conventional levels (stage 3, interpersonal accord and conformity; stage 4, social accord and system maintenance); and a postconventional level (stage 5, social contract, utility, individual rights). Studies suggest that by age thirty-six, 89 percent of middle-class American males have developed to the conventional stage of moral reasoning and 11 percent to the postconventional stage. Elliot attained a global score of 4/5, indicating a late-conventional, early-postconventional mode of moral thought. This is an excellent result.

In brief, Elliot had a normal ability to generate response options to social situations and to consider spontaneously the consequences of particular response options. He also had a capacity to conceptualize means to achieve social objectives, to predict the likely outcome of social situations, and to perform moral reasoning at an advanced developmental level. The findings indicated clearly that damage to the ventromedial sector of the frontal lobe did not destroy the records of social knowledge as retrieved under the conditions of the experiment.[7]

While Elliot's preserved performance was consonant with his superior scoring on conventional tests of memory and intellect, it contrasted sharply with the defective decision-making he exhibited in real life. How could this be explained? We accounted for the dramatic dissociation on the basis of several differences between the conditions and demands of these tasks and the conditions and demands of real life. Let us analyze those differences.

Except for the last task, there was no requirement to make a choice among options. It was sufficient to conjure up options and likely consequences. In other words, it was sufficient to reason through the problem, but not necessary for reasoning to abut a decision. Normal performance in this task demonstrated the existence of social knowledge and access to it, but said nothing about the process or choice itself. Real life has a way of forcing you into choices. If you do not succumb to the forcing, you can be just as undecided as Elliot.

The above distinction is illustrated best in Elliot's own words. At the end of one session, after he had produced an abundant quantity of options for action, all of which were valid and implementable, Elliot smiled, apparently satisfied with his rich imagination, but added: "And after all this, I still wouldn't know what to do!"

Even if we had used tests that required Elliot to make a choice on every item, the conditions still would have differed from real-life circumstances; he would have been dealing only with the original set of constraints, and not with new constraints resulting from an initial response. If it had been "real life," for every option Elliot offered in a given situation there would have been a response from the other side, which would have changed the situation and required an additional set of options from Elliot, which would have led to yet another response, and in turn to another set of options required from him, and so on. In other words, the ongoing, open-ended, uncertain evolution of real-life situations was missing from the laboratory tasks. The purpose of Jeffrey Saver's study, however, was to assess the status and accessibility of the knowledge base itself, not the reasoning and deciding process.

I should point out other differences between real life and the laboratory tasks. The time frame of the events under consideration in the tasks was compacted rather than real. In some circumstances, real-time processing may require holding information—representations of persons, objects, or scenes, for instance—in mind for longer periods, especially if new options or consequences surface and require comparison. Furthermore, in our tasks, the situations and questions about them were presented almost entirely through language. More often than not, real life faces us with a greater mix of pictorial and linguistic material. We are confronted with people and objects; with sights, sounds, smells, and so on; with scenes of varying intensities; and with whatever narratives, verbal and or pictorial, we create to accompany them.

These shortcomings aside, we had made progress. The results strongly suggested that we should not attribute Elliot's decision-making defect to lack of social knowledge, or to deficient access to such knowledge, or to an elementary impairment of reasoning, or, even less, to an elementary defect in attention or working memory concerning the processing of the factual knowledge needed to make decisions in the personal and social domains. The defect appeared to set in at the late stages of reasoning, close to or at the point at which choice making or response selection must occur. In other words, whatever went wrong went wrong late in the process. Elliot was unable to choose effectively, or he might not choose at all, or choose badly. Remember how he would drift from a given task and spend hours sidetracked? As we are confronted by a task, a number of options open themselves in front of us and we must select our path correctly, time after time, if we are to keep on target. Elliot could no longer select that path. Why he could not is what we needed to discover.

I was now certain that Elliot had a lot in common with Phineas Gage. Their social behavior and decision-making defect were compatible with a normal social-knowledge base, and with preserved higher-order neuropsychological functions such as conventional memory, language, basic attention, basic working memory and basic reasoning. Moreover, I was certain that in Elliot the defect was accompanied by a reduction in emotional reactivity and feeling. (In all likelihood the emotional defect was also present in Gage, but the record does not allow us to be certain. We can infer at least that he lacked the feeling of embarrassment, given his use of foul language and his parading of self-misery.) I also had a strong suspicion that the defect in emotion and feeling was not an innocent bystander next to the defect in social behavior. Troubled emotions probably contributed to the problem. I began to think that the cold-bloodedness of Elliot's reasoning prevented him from assigning different values to different options, and made his decision-making landscape hopelessly flat. It might also be that the same cold-bloodedness made his mental landscape too shifty and unsustained for the time required to make response selections, in other words, a subtle rather than basic defect in working memory which might alter the remainder of the reasoning process required for a decision to emerge. Be that as it may, the attempt to understand both Elliot and Gage promised an entry into the neurobiology of rationality.

Four

In Colder Blood

THERE NEVER HAS been any doubt that, under certain circumstances, emotion disrupts reasoning. The evidence is abundant and constitutes the source for the sound advice with which we have been brought up. Keep a cool head, hold emotions at bay! Do not let your passions interfere with your judgment. As a result, we usually conceive of emotion as a supernumerary mental faculty, an unsolicited, nature-ordained accompaniment to our rational thinking. If emotion is pleasurable, we enjoy it as a luxury; if it is painful, we suffer it as an unwelcome intrusion. In either case, the sage will advise us. we should experience emotion and feeling in only judicious amounts. We should be reasonable.

There is much wisdom in this widely held belief, and I will not deny that uncontrolled or misdirected emotion can be a major source of irrational behavior. Nor will I deny that seemingly normal reason can be disturbed by subtle biases rooted in emotion. For instance, a patient is more likely to prefer a treatment if told that 90 percent of those treated are alive five years later, than if told that 10 percent are dead.[1] Although the outcome is precisely the same, it is likely that the feelings aroused by the idea of death lead to the rejection of an option that would be endorsed in the other framing of the choice, in short, an inconsistent and irrational inference. That the irrationality does not result from lack of knowledge is borne out by the fact that physicians respond no differently than non-physician patients. Nonetheless, what the traditional account leaves out is a notion that emerges from the study of patients such as Elliot and from other observations I discuss below: *Reduction in emotion may constitute an equally important source of irrational behavior.* The counterintuitive connection between absent emotion and warped behavior may tell us something about the biological machinery of reason.

I began pursuing this notion utilizing the approach of experimental neuropsychology.[2] Roughly, the approach depends on the following steps: finding systematic correlations between damage at given brain sites and disturbances of behavior and cognition; validating the findings by establishing what are known as double dissociations, in which damage at site A causes disturbance X but not disturbance Y, while damage at site B causes disturbance Y but not disturbance X; formulating both general and particular hypotheses according to which a normal neural system made up of different components (e.g., cortical regions and subcortical nuclei) performs a normal cognitive/behavioral operation with different fine-grain components; and finally, testing the hypotheses in new cases of brain damage in which a lesion at a given site is used as a *probe* to whether damage has caused the hypothesized effect.

The goal of the neuropsychological enterprise is thus to explain how certain cognitive operations and their components relate to neural systems and their components. Neuropsychology is not, or should not be, about finding the brain "localization" for a "symptom" or "syndrome."

My first concern was to verify that our observations about Elliot held firm in other patients. That proved to be the case. To date we have studied twelve patients with prefrontal damage of the type seen in Elliot, and in none have we failed to encounter a combination of

decision-making defect and flat emotion and feeling. The powers of
reason and the experience of emotion decline together, and their impairment stands out in a neuropsychological profile within which
basic attention, memory, intelligence, and language appear so intact
that they could never be invoked to explain the patients' failures in
judgment.

But the salient, concurrent impairment of reason and feeling does
not arise only after prefrontal damage. In this chapter, I will show
how this combination of impairments can arise from damage to
other specific brain sites and how such correlations suggest an
interaction of the systems underlying the normal processes of emotion, feeling, reason, and decision making.

EVIDENCE FROM OTHER
CASES OF PREFRONTAL DAMAGE

I should place my comments about cases of prefrontal damage in a
historical perspective. Phineas Gage's case is not the only important
historical source in the effort to understand the neural basis of
reasoning and decision making; I can offer four other sources to help
round out the basic profile.

The first, studied in 1932 by Brickner, a neurologist at Columbia
University, and identified as "patient A," was a thirty-nine-year-old
New York stockbroker, personally and professionally successful, who
developed a brain tumor, like Elliot's a meningioma.[3] The tumor
grew from above and pressed down on the frontal lobes. The result
was similar to what we saw in Elliot. The pioneer neurosurgeon
Walter Dandy was able to remove the life-threatening tumor but not
before the mass had done extensive damage to the cerebral cortices
in the frontal lobes, on the left and on the right. The affected areas
included all those that were lost in Elliot and in Gage, and went a bit
beyond. On the left, all the frontal cortices located in front of the
areas for language were removed. On the right, the excision was
larger and included all the cortex in front of the areas controlling
movement. The cortices in the ventral (orbital) surface and the lower

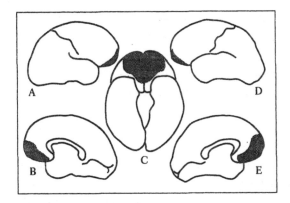

Figure 4-1. The shaded areas represent the ventral and medial sectors of the frontal lobe which are consistently compromised in patients with the "Gage matrix." Note that the dorsolateral sector of the frontal lobes is not affected.

    A: Right cerebral hemisphere, external (lateral) view.

    B: Right cerebral hemisphere, internal (medial) view.

    C: The brain viewed from below (ventral or orbital view).

    D: Left hemisphere, external view.

    E: Left hemisphere, internal view.

part of the internal (medial) surface of both sides of the frontal lobes were also removed. The cingulate was spared. (The entire surgical description was confirmed twenty years later, at autopsy).

Patient A had normal perception. His orientation to person, place, and time was normal, as was his conventional memory for recent and remote facts. His language and motor abilities were unaffected, and his intelligence seemed intact, on the basis of the psychological tests available at the time. Much was made of the fact that he could perform calculations and play a good game of checkers. But in spite of his impressive physical health and commendable mental abilities, patient A never returned to work. He stayed home, formulating plans for his professional comeback but never implementing the simplest of those plans. Here was another life unraveling.

A's personality had changed profoundly. His former modesty had vanished. He had been polite and considerate, but now he could be embarrassingly inappropriate. His remarks about other people, including his wife, were uncaring and sometimes downright cruel. He boasted of his professional, physical, and sexual prowess, although he did not work, did nothing sporty, and had stopped having sex with his wife or anyone else. Much of his conversation revolved around mythical exploits and was peppered by facetious remarks, generally at the expense of others. On occasion, if frustrated, he would be verbally abusive though never physically violent.

Patient A's emotional life seemed impoverished. Now and then he might have a short-lived burst of emotion, but for the most part such display was lacking. There is no sign that he felt for others, and no sign of embarrassment, sadness, or anguish at such a tragic turn of events. His overall affect is best captured as "shallow." By and large, patient A had become passive and dependent. He spent the rest of his life under the supervision of his family. He was taught to operate a printing machine on which he made visiting cards, and that became his only productive activity.

Patient A clearly exhibited the cognitive and behavioral characteristics I am trying to establish for what one might call the Phineas Gage matrix: after he sustained damage to the frontal cortices, his ability to choose the most advantageous course of action was lost, despite otherwise intact mental capacities; emotions and feelings were compromised. Around this matrix, to be sure, there are differences in personality profile when several cases are compared. But it is in the inevitable nature of syndromes to have a matrix, a shared essence of symptoms, and to have symptom variance around the edges of that essence. As I indicated in discussing the surface differences between Gage and Elliot, it is premature to decide on the cause of those differences. At this point I want merely to emphasize the shared essence of the condition.

The second historical source dates from 1940.[4] Donald Hebb and Wilder Penfield, at McGill University in Canada, described a patient who had been in a serious accident at age sixteen, and they addressed an important point. Phineas Gage, patient A, and their modern counterparts had been normal adults and had attained a mature personality before they suffered damage to the frontal lobes and showed signs of abnormal behavior. What if the damage had occurred during development, sometime in childhood or adolescence? One might predict that children or adolescents so impaired would never develop a normal personality, that their social sense would never mature, and that is precisely what has been found in such cases. The Hebb–Penfield patient had a compound fracture of the frontal bones which compressed and destroyed the frontal cortices on both sides. He had been a normal child and a normal adolescent; after the injury, however, not only was his continued social development arrested, but his social behavior deteriorated.

Perhaps even more telling is the third case, described by S. S. Ackerly and A. L. Benton in 1948.[5] Their patient sustained frontal lobe damage around the time of birth and thus went through childhood and adolescence without many of the brain systems that I believe are necessary for a normal human personality to emerge. Accordingly, his behavior was always abnormal. Although he was not a stupid child, and although the basic instruments of his mind seemed intact, he never acquired normal social behavior. When a neurosurgical exploration was performed at age nineteen, it revealed that the left frontal lobe was little more than a hollow cavity and the entire right frontal lobe was absent as a consequence of atrophy. Severe damage at about the time of birth had irrevocably damaged most of the frontal cortices.

This patient was never able to hold a job. After some days of obedience he would lose interest in his activity, and even end up stealing or being disorderly. Any departure from routine would frustrate him easily and might cause a burst of bad temper, although in general he tended to be docile and polite. (He was described as having the courteous manner known as "English valet politeness.") His sexual interests were dim, and he never had an emotional involvement with any partner. His behavior was stereotyped, unimaginative, lacking in initiative, and he developed no professional skills or hobbies. Reward or punishment did not seem to influence his behavior. His memory was capricious; it failed in instances in which one would expect learning to occur, and suddenly might succeed on some peripheral subject, e.g., a detailed knowledge of the makes of automobiles. The patient was neither happy nor sad, and his pleasure and pain both seemed short-lived.

The Hebb–Penfield and Ackerly–Benton patients shared a number of personality traits. Rigid and perseverant in their approach to life, they both were unable to organize future activity and hold gainful employment; they lacked originality and creativity; they tended to boast and present a favorable view of themselves; they displayed generally correct but stereotyped manners; they were less able than others to experience pleasure and react to pain; they had diminished sexual and exploratory drives; and they demonstrated a lack of motor, sensory, or communication defects, and an overall intelligence within expectations, given their sociocultural background. Modern counterparts of such cases continue to present themselves, and in those I have observed, the consequences are similar. The patients resemble Ackerly and Benton's in medical history and social behavior. One way of describing their predicament is by saying that they never construct an appropriate theory about their persons, or about their person's social role in the perspective of past and future. And what they cannot construct for themselves, they also cannot generate for others. They are bereft of a theory of their own mind and of the mind of those with whom they interact.[6]

The fourth source of historical evidence is from an unexpected quarter: the literature on prefrontal leucotomy. This surgical procedure, developed in 1936 by the Portuguese neurologist Egas Moniz, was meant to treat the anxiety and agitation accompanying psychiatric conditions such as obsessive-compulsive disease and schizophrenia.[7] As originally designed by Moniz and carried out by his collaborator, the neurosurgeon Almeida Lima, the surgery produced small areas of damage in the deep white matter of both frontal lobes. (The name of the procedure is simple enough: *leukos* is Greek for "white," and *tomos* is Greek for "section"; "prefrontal" indicates the region targeted in the operation.) As was discussed in chapter 2, the white matter below the cerebral cortex is made up of bundles of axons, or nerve fibers, each of which is a prolongation of a neuron. The axon is the means by which one neuron makes contact with another. The bundles of axons crisscross the brain substance in the white matter, connecting different regions of the cerebral cortex. Some connections are local, between regions of cortex just a few millimeters away from each other, while other connections link regions that are farther apart, for instance, cortical regions in one cerebral hemisphere to cortical regions in the other. There are also connections in one direction or the other between cortical regions and subcortical nuclei, the aggregates of neurons below the cerebral cortex. A bundle of axons from a known source to a given target is often referred to as a "projection," because the axons project to a particular collection of neurons. A sequence of projections across several target stations is known as a "pathway."

The novel idea Moniz had conceived was that in patients with pathologic anxiety and agitation, projections and pathways of white matter in the frontal region had established abnormally repetitive and overactive circuits. There was no evidence for such a hypothesis, although recent studies on the activity of the orbital region in obsessive and depressed patients suggest that Moniz may have been correct, at least in part, even where the details were wrong. But if Moniz's idea was bold and ahead of the evidence at the time, it was almost timid compared with the treatment he would propose. Reasoning from the case of patient A, and from the results of animal experiments to be discussed below, Moniz predicted that a surgical severing of those connections would abolish anxiety and agitation while leaving intellectual capacities undisturbed. He believed such an operation would cure the patients' suffering and permit them to lead a normal mental life. Motivated by what he saw as the desperate state of so many untreated patients, Moniz developed and attempted the operation.

The results of the initial prefrontal leucotomies gave some support to Moniz's predictions. The patients' anxiety and agitation were abolished, and functions such as language and conventional memory remained largely intact. It would not be correct, however, to assume that the surgery did not impair the patients in other ways. Their behavior, which had never been normal, was now abnormal in a different manner. Extreme anxiety gave way to extreme calm. Their emotions seemed flat. They did not appear to suffer. The animated intellect which had produced incessant compulsions or rich delusions was quiet. The patients' drive to respond and act, however wrongly, was muffled.

The evidence from these early procedures is far from ideal. It was collected long ago, with the limited neuropsychological knowledge and instruments of the time, and it is not as free of prejudices, positive or negative, as one would wish. The controversy over this modality of treatment was overwhelming. Yet the existing studies do point to the following facts: First, damage to the white matter subjacent to the orbital and medial regions of the frontal lobe altered emotion and feeling, drastically reducing both. Second, the basic instruments of perception, memory, language, and movement were not affected. And third, to the degree that it is possible to separate new behavioral signs from those that led to the intervention, it appears that leucotomized patients were less creative and decisive than before.

In fairness to Moniz and to the early prefrontal leucotomy procedure, it should be noted that unquestionably the patients drew some benefit from the surgery. An additional degree of decision-making defect, in the background of their primary psychiatric illness, was perhaps a smaller burden to bear than their uncontrolled anxiety had been. Much as a surgical mutilation of the brain is unacceptable, we must remember that in the 1930s, typical treatment for such patients involved committing them to mental institutions and/or administering massive doses of sedatives which only blunted their anxiety when they were virtually stunned into sleep. The few alternatives to leucotomy included the straitjacket and shock therapy. Not until the late 1950s did psychotropic drugs such as Thorazine begin to appear. We must remember also that we still have no way of knowing whether the long-term effects of such drugs on the brain are any less destructive than a selective form of surgery might be. We simply have to reserve judgment.

There is no need, though, to reserve judgment against the far more destructive version of Moniz's intervention known as frontal lobotomy. The operation conceived by Moniz caused limited brain damage. Frontal lobotomy, in contrast, was often a butchering affair which caused extensive lesions. It became infamous worldwide, for the questionable way in which it was prescribed and for the unnecessary mutilation it produced.[8]

On the basis of the historical documentation and of the evidence obtained in our laboratory, we reached the following provisional conclusions:

1. If the ventromedial sector is included in the lesion, bilateral damage to prefrontal cortices is consistently associated with impairments of reasoning/decision making and emotion/feeling.

2. When impairments in reasoning/decision making and emotion/feeling stand out against an otherwise largely intact neuropsychological profile, the damage is most extensive in the ventromedial sector; moreover the personal/social domain is the one most affected.

3. In cases of prefrontal damage in which the dorsal and lateral sectors are damaged at least as extensively as the ventromedial sector if not more so, impairments in reasoning/decision making are no longer concentrated in the personal/social domain. Those impairments, as well as the impairments in emotion/feeling are accompanied by defects in attention and working memory detected by tests in which objects, words, or numbers are used.

What we needed to know now was whether the strange bedfellows—impaired reasoning/decision making and impaired emotion/feeling—could show up alone or in other neuropsychological company, as a result of damage elsewhere in the brain.

The answer was that they could. They showed up prominently as a result of damage in other sites. One of these was a sector of the right (but not left) cerebral hemisphere that contains the several cortices in charge of processing signals from the body. Another included structures of the limbic system such as the amygdala.

EVIDENCE FROM DAMAGE
BEYOND PREFRONTAL CORTICES

There is another important neurological condition that shares the Phineas Gage matrix, even if affected patients do not resemble Gage on the surface. Anosognosia, as the condition is known, is one of the most eccentric neuropsychological presentations one is likely to encounter. The word—which derives from the Greek *nosos*, "disease," and *gnosis*, "knowledge"—denotes the inability to acknowledge disease in oneself. Imagine a victim of a major stroke, entirely paralyzed in the left side of the body, unable to move hand and arm, leg and foot, face half immobile, unable to stand or walk. And now imagine that same person oblivious to the entire problem, reporting that nothing is possibly the matter, answering the question, "How do you *feel?*" with a sincere, "Fine." (The term *anosognosia* has been used also to designate unawareness of blindness or aphasia. In my discussion I refer only to the prototypical form of the condition, as noted above and first described by Babinski.[9])

Someone unacquainted with anosognosia might think that this "denial" of illness is "psychologically" motivated, that it is an adaptive reaction to the previous affliction. I can state with confidence that this is not the case. Consider the mirror image tragedy, the one in which the *right* side of the body is paralyzed rather than the left: patients so affected usually do not have anosognosia, and although they are often severely incapacitated in their use of language and may suffer from aphasia, they are fully cognizant of their plight. Furthermore, some patients who have a devastating left-side paralysis, but caused by a pattern of brain damage different from the one that causes paralysis *and* anosognosia, can be normal in their mind and behavior and realize their handicap. In short, left-side paralysis caused by a particular pattern of brain damage is accompanied by anosognosia; right-side paralysis caused by the mirror-image pattern of brain damage is not accompanied by anosognosia; left-side paralysis caused by patterns of brain damage other than those associated with anosognosia is not accompanied by unawareness. Anosognosia, then, occurs systematically with damage to a particular region of the brain, and only that region, in patients who may appear, to people unfamiliar with neurological mystery, more fortunate than those who are both half paralyzed and language-impaired. The "denial" of illness results from the loss of a particular cognitive function. This loss of cognitive function depends on a particular brain system which can be damaged by a stroke or by various neurological diseases.

Typical anosognosics need to be confronted with their blatant defect so that they will know there is something the matter with them. Whenever I asked my patient DJ about her left-side paralysis, which was complete, she would always begin by saying that her movements were entirely normal, that perhaps they had once been impaired but they no longer were. When I would ask her to move her left arm, she would search around for it and, after looking at the inert limb, ask whether I really wanted "it" to move "by itself." When I would say yes, please, she would then take *visual* notice of the lack of any motion in the arm, and tell me that "it doesn't seem to do much by itself." As a sign of cooperation, she would offer to have the good hand move the bad arm: "I can move it with my right hand."

This inability to sense the defect automatically, rapidly, and internally, through the body's sensory system, never disappears in severe cases of anosognosia, although in mild cases it can be masked. For instance, a patient may have the visual recollection of the motionless limb and by inference realize that something is the matter with that part of the body. Or a patient may recall the countless statements, from relatives and medical staff, to the effect that there is paralysis, there is disease, that no, things are not normal. Relying on that sort of extraneously obtained information, one of our most intelligent anosognosics consistently says, "I *used* to have that problem," or, "I *used* to have neglect." Of course, he still does. The lack of direct update on the real state of body and person is nothing less than astounding. (Unfortunately, this subtle distinction between patients' direct and indirect awareness of their condition is often missed or glossed over in discussions of anosognosia. For a rare exception see A. Marcel.[10])

No less dramatic than the oblivion that anosognosic patients have regarding their sick limbs is the lack of concern they show for their overall situation, the lack of emotion they exhibit, the lack of feeling they report when questioned about it. The news that there was a major stroke, that the risk of further trouble in brain or heart looms large, or the news that they are suffering from an invasive cancer that has now spread to the brain—in short, the news that life is not likely to be the same, ever again—is usually received with equanimity, sometimes with gallows humor, but never with anguish or sadness, tears or anger, despair or panic. It is important to realize that if you give a comparable set of bad news to a patient with the mirror image damage in the left hemisphere the reaction is entirely normal. Emotion and feeling are nowhere to be found in anosognosic patients, and perhaps this is the only felicitous aspect of their otherwise tragic condition. Perhaps it is no surprise that these patients' planning for the future, their personal and social decision-making, is profoundly impaired. Paralysis is perhaps the least of their troubles.

In a systematic study of anosognosic patients, the neuropsychologist Steven Anderson has confirmed the extensive range of defects and demonstrated that the patients are as neglectful of their situation and of its consequences as they are of their paralysis.[11] Many appear unable to foresee the likelihood of dire consequences; if and when they do predict them, they appear unable to suffer accordingly. They certainly cannot construct an adequate theory for what is happening to them, for what may happen in the future, and for what others think of them. Just as important, they are unaware that their own theorizing is inadequate. When one's own self-image is so compromised, it may not be possible to realize that the thoughts and actions of that self are no longer normal.

. . .

Patients with the type of anosognosia described above have damage in the right hemisphere. Although drawing up a full characterization of the neuroanatomical correlates of anosognosia is an ongoing project, this much is apparent: There is damage to a select group of right cerebral cortices which are known as somatosensory (from the Greek root *soma*, for body; the somatosensory system is responsible for both the external senses of touch, temperature, pain, and the internal senses of joint position, visceral state, and pain) and which include the cortices in the insula; the cytoarchitectonic areas 3, 1, 2 (in the parietal region); and area S2 (also parietal, in the depth of the sylvian fissure). (Note that whenever I use the term somatic or somatosensory I have in mind the soma, or body, in the general sense, and I refer to all types of body sensation including visceral sensations.) The damage also affects the white matter of the right hemisphere, disrupting the interconnection among the above-mentioned regions, which receive signals from throughout the body (muscles, joints, internal organs), and their interconnection with the thalamus, the basal ganglia, and the motor and prefrontal

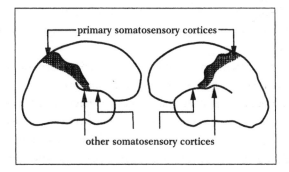

Figure 4-2. Diagram of a human brain showing the right and left hemispheres seen from the outside. The shaded areas cover the primary somatosensory cortices. Other somatosensory areas, respectively the second sensory area (S2) and the insula, are buried inside the sylvian fissure immediately anterior and posterior to the bottom of the primary somatosensory cortex. They are thus not visible in a surface rendering. Their approximate location in the depth is identified by the arrows.

cortices. Partial damage to the multicomponent system described here, does *not* cause the type of anosognosia I am discussing.

It has long been my working assumption that the brain areas that cross-talk within the overall region of the right hemisphere damaged in anosognosia, probably produce, through their cooperative inter-actions, the most comprehensive and integrated map of the current body state available to the brain.

The reader may wonder why this map is skewed to the right hemisphere rather than being bilateral; after all, the body has two almost symmetrical halves. The answer is that in human as well as nonhuman species, functions seem to be apportioned *a*symmetrically to the cerebral hemispheres, for reasons which probably have to do with the need for one final controller rather than two, when it comes to choosing an action or a thought. If both sides had equal say on making a movement, you might end up with a conflict—your right hand might interfere with the left, and you would have a lesser chance of producing coordinated patterns of motion involving more than one limb. For a variety of functions, structures in one hemisphere must have an advantage; those structures are called *dominant*.

The best-known example of dominance concerns language. In more than 95 percent of all people, including many left-handers, language depends largely on left-hemisphere structures. Another example of dominance, this one favoring the right hemisphere, involves integrated body sense, through which the representation of visceral states, on the one hand, and the representation of states of limb, trunk, and head components of the musculoskeletal appa-ratus, on the other, come together in a coordinated dynamic map. Note that this is not a single, contiguous map, but rather an interac-tion and coordination of signals in separate maps. In this arrange-ment, signals concerning both left and right sides of the body find their most comprehensive meeting ground in the right hemisphere in the three somatosensory cortical sectors indicated previously. Intriguingly, the representation of extrapersonal space, as well as the processes of emotion, involve a right-hemisphere dominance.[12] This is not to say that the equivalent structures in the left hemisphere do not represent the body, or space for that matter. It is just that the representations are different: left-hemisphere representations are probably partial and not integrated.

Patients with anosognosia resemble those with prefrontal damage, in some respects. Anosognosics, for instance, are unable to make appropriate decisions on personal and social matters, just as is the case with prefrontal patients. And prefrontal patients with decision-making impairment are, like anosognosics, usually indifferent to their health status and seem to have an unusual tolerance for pain.

Some readers may be surprised at this, and may ask why they haven't heard more about the decision-making impairments of anosognosics. Why has the little interest accorded to impaired reasoning after brain damage been centered on prefrontally damaged patients? We might consider, by way of explanation, that patients with prefrontal lesions appear neurologically normal (their movements, sensations, and language are intact; the disturbance resides with their impaired feelings and reasonings) and thus can engage in a variety of social interactions that will easily expose their defective reasoning. Patients with anosognosia, on the other hand, are more often than not considered sick, because of their blatant motor and sensory impairments, and are thus limited in the range of social interactions in which they can engage. In other words, their opportunity to place themselves in harm's way is drastically reduced. Even so, the decision-making defects are there, ready to manifest themselves given the opportunity, ready to undermine the best rehabilitation plans made for such patients by families and medical staff. Unable to realize how profoundly impaired they are, these patients show little or no inclination to cooperate with therapists, no motivation at all to get better. Why should they, if they are generally unaware of how badly off they are in the first place? The appearance of cheerfulness or indifference is deceptive, since such appearances are not voluntary and are not based on knowledge of the situation. Yet these appearances often are misinterpreted as adaptive, and caregivers are misled into giving a better prognosis for outwardly cheerful patients than for their teary, anguished counterparts next door.

A pertinent example in this regard is that of Supreme Court Justice William O. Douglas, who in 1975 suffered a right-hemisphere stroke.[13] The lack of language defects augured well for his return to the bench, or so people thought, hoping that this brilliant and decisive member of the Court would not be lost prematurely. But the sad events that followed told a different story, and show how the consequences may be problematic when a patient with these impairments is allowed to have extensive social interactions.

The telltale signs came early, when Douglas checked himself out of the hospital against medical advice (he would do this more than once, and have himself driven to the Court, or on exhausting shopping and dining sprees). This, as well as the jocular way with which he attributed his hospitalization to a "fall," and dismissed the left-side paralysis as a myth, was attributed to his proverbial firmness and humor. When he was forced to realize and admit, in an open press conference, that he could not walk or get out of his wheelchair unaided, he dismissed the matter by saying, "Walking has very little to do with the work of the Court." Nonetheless, he invited reporters to go hiking with him the following month. Later, after renewed efforts at rehabilitation had proved fruitless, Douglas replied to a visitor who asked about his left leg, "I've been kicking forty-yard field goals with it in the exercise room," and ventured that he would sign up with the Washington Redskins. When the stunned visitor politely countered that his advanced age might put a damper on the project, the justice laughed and said, "Yes, but you ought to see how I'm arching them." The worst was yet to come, though, as Douglas repeatedly failed to observe social convention with the other justices and staff. Although unable to perform his job, he steadfastly refused to resign, and even after he was forced to do so, he often behaved as if he had not.

Anosognosics of the type I described here, then, have more than just a left-side paralysis of which they are not aware. They also have a

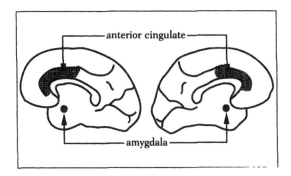

Figure 4-3. Looking at the internal surface of both hemispheres. The shaded areas cover the anterior cingulate cortex. The black disk marks the projection of the amygdala onto the internal surface of the temporal lobes.

defect in reasoning and decision making, and a defect in emotion and feeling.

Now a word about evidence from damage to the amygdala, one of the most important components of the limbic system. Patients with bilateral damage confined to the amygdala are exceedingly rare. My colleagues Daniel Tranel, Hanna Damasio, Frederick Nahm, and Bradley Hyman have been fortunate to study one such patient, a woman with a lifelong pattern of personal and social inadequacy.[14] There is no doubt that the range and appropriateness of her emotions are impaired and that she has little concern for the problematic situations into which she gets herself. The "folly" of her behavior is not unlike that found in Phineas Gage or patients with anosognosia, and, as in them, it cannot be blamed on poor education or low intelligence (the woman in question is a high school graduate, and her IQ is in the normal range). Moreover, in a series of ingenious experiments, Ralph Adolphs has shown that this patient's appreciation of subtle aspects of emotion is profoundly abnormal. Although these findings must be replicated in comparable cases before too much weight is placed on them, I must add that equivalent lesions in monkeys cause a defect in emotional processing, as first shown by

Larry Weiskrantz and confirmed by Aggleton and Passingham.[15] Furthermore, working in rats, Joseph LeDoux has shown beyond the shadow of a doubt that the amygdala plays a role in emotion (more about this finding in chapter 7).[15]

A REFLECTION ON ANATOMY AND FUNCTION

The preceding survey of neurological conditions in which impairments of reasoning/decision making, and emotion/feeling figure prominently reveals the following:

First, there is a region of the human brain, the ventromedial prefrontal cortices, whose damage consistently compromises, in as pure a fashion as one is likely to find, both reasoning/decision making, and emotion/feeling, especially in the personal and social domain. One might say, metaphorically, that reason and emotion "intersect" in the ventromedial prefrontal cortices, and that they also intersect in the amygdala.

Second, there is a region of the human brain, the complex of somatosensory cortices in the right hemisphere, whose damage also compromises reasoning/decision making and emotion/feeling, and, in addition, disrupts the processes of basic body signaling.

Third, there are regions located in prefrontal cortices beyond the ventromedial sector, whose damage also compromises reasoning and decision making, but in a different pattern: Either the defect is far more sweeping, compromising intellectual operations over all domains, or the defect is more selective, compromising operations on words, numbers, objects, or space, more so than operations in the personal and social domain. A rough map of these critical intersections is shown in Figure 4-4.

In short, there appears to be a collection of systems in the human brain consistently dedicated to the goal-oriented thinking process we call reasoning, and to the response selection we call decision making, with a special emphasis on the personal and social domain. This same collection of systems is also involved in emotion and feeling, and is partly dedicated to processing body signals.

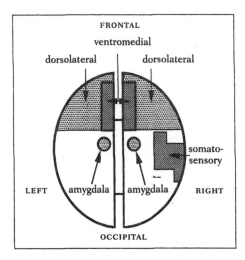

FRONTAL
ventromedial
dorsolateral   dorsolateral

somato-
sensory

LEFT   amygdala   amygdala   RIGHT

OCCIPITAL

Figure 4-4. A diagram representing the set of regions whose damage compromises both aspects of reasoning and processing of emotion.

A FOUNTAINHEAD

Before leaving the subject of human brain lesions, I would like to propose that there is a particular region in the human brain where the systems concerned with emotion/feeling, attention, and working memory interact so intimately that they constitute the source for the energy of both external action (movement) and internal action (thought animation, reasoning). This fountainhead region is the anterior cingulate cortex, another piece of the limbic system puzzle.

My idea about this region comes from observing a group of patients with damage in and around it. Their condition is described best as suspended animation, mental and external—the extreme variety of an impairment of reasoning and emotional expression. Key regions affected by the damage include the anterior cingulate cortex (I may refer to it simply as "cingulate"), the supplementary motor area (the latter is known as SMA or M2), and the third motor area (known as M3).[16] In some cases, adjoining prefrontal areas are involved too, as may be the motor cortex in the inner surface of the hemisphere. As a whole, the areas contained in this sector of the frontal lobe have been associated with movement, emotion, and attention. (Their involvement in motor function is well established; for evidence on their involvement in emotion and attention, see Damasio and Van Hoesen, 1983, and Petersen and Posner, 1990, respectively.[17]) Damage to this sector not only produces impairment in movement, emotion, and attentiveness, but also causes a virtual suspension of the animation of action and of thought process such that reason is no longer viable. The story of one of my patients in whom there was such damage gives an idea of the impairment.

The stroke suffered by this patient, whom I will call Mrs. T, produced extensive damage to the dorsal and medial regions of the frontal lobe in both hemispheres. She suddenly became motionless and speechless, and she would lie in bed with her eyes open but with a blank facial expression; I have often used the term "neutral" to convey the equanimity—or absence—of such an expression.

Her body was no more animated than her face. She might make a normal movement with arm and hand, to pull her bed covers for instance, but in general, her limbs were in repose. When asked about

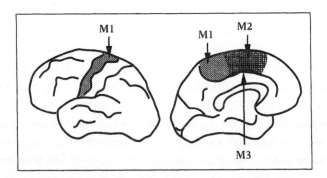

Figure 4-5. Diagram of the human brain representing the left cerebral hemisphere seen from the outside (left panel) and the inside (right panel). The location of the three main cortical motor regions: M1, M2, and M3. M1 includes the so-called "motor strip" which shows up in every cartoon of the brain. An ugly human figure ("Penfield's homunculus") is often drawn on top of it. The less well known M2 is the supplementary motor area, the internal part of area 6. Even less known is M3 which is buried in the depth of the cingulate sulcus.

her situation, she usually would remain silent, although after much coaxing she might say her name, or the names of her husband and children, or the name of the town where she lived. But she would not tell you about her medical history, past or present, and she could not describe the events leading to her admission to the hospital. There was no way of knowing, then, whether she had no recollection of those events or whether she had a recollection but was unwilling or unable to talk about it. She never became upset with my insistent questioning, never showed a flicker of worry about herself or anything else. Months later, as she gradually emerged from this state of mutism and akinesia (lack of movement), and began to answer questions, she would clarify the mystery of her state of mind. Contrary to what one might have thought, her mind had not been imprisoned in the jail of her immobility. Instead it appeared that there had not been much mind at all, no real thinking or reasoning. The passivity in her face and body was the appropriate reflection of her lack of mental animation. At this later date she was certain about not having felt anguished by the absence of communication. Nothing had forced her not to speak her mind. Rather, as she recalled, "I really had nothing to say."

To my eyes Mrs. T had been unemotional. To her experience, all the while, it appears she had had no feelings. To my eyes she had not specifically attended to the external stimuli presented to her, nor had she attended internally to their representation or to the representation of correlated evocations. I would say her will had been preempted, and that seems also to have been her reflection. (Francis Crick has drawn on my suggestion that volition was preempted in patients with such lesions, and discussed a neural substrate for free will.[18]) In short, there was a pervasive impairment of the drive with which mental images and movements can be generated and of the means by which they can be enhanced. The lack of that drive was translated externally to a neutral facial expression, mutism, and akinesia. It appears that there had been no normally differentiated thought and reasoning in Mrs. T's mind, and naturally no decisions made and even less implemented.

EVIDENCE FROM ANIMAL STUDIES

Further background for the argument I am constructing comes from animal studies. The first study I will discuss dates back to the 1930s. An observation made in chimpanzees seems to have been if not the spark for the prefrontal leucotomy project, at least the strong encouragement Moniz needed to proceed with his idea. The observation was made by J. F. Fulton and C. F. Jacobsen at Yale University, in the course of studies aimed at understanding learning and memory.[19] Becky and Lucy, two chimpanzees they were working with, were not pleasant creatures; when they were frustrated, as they easily were, they became vicious. In the course of the study, Fulton and Jacobsen wanted to investigate how damage to the prefrontal cortex would alter the animals' learning of an experimental task. In a first stage, the researchers damaged one frontal lobe. Nothing much happened to the performance or to the animals' personalities. In the next stage, the researchers damaged the other frontal lobe. And then something remarkable did happen. In circumstances in which Becky and Lucy previously had been frustrated, they now seemed not to mind; instead of being vicious they now were placid. Jacobsen described the transformation in vivid terms to a roomful of colleagues in London during the 1935 World Congress of Neurology.[20] Upon hearing his remarks, Moniz is supposed to have stood up and asked whether similar lesions made in the brains of psychotic patients would not provide a solution to some of their problems. A startled Fulton was unable to answer.

Bilateral prefrontal damage as described above precludes normal emotional display and, no less important, causes abnormalities in social behavior. In a series of revealing studies, Ronald Myers has shown that monkeys with bilateral prefrontal ablations (involving both the ventromedial and the dorsolateral sectors but sparing the cingulate region) do not maintain normal social relations within the monkey troop despite the fact that nothing in their physical appearance has changed.[21] These affected monkeys show greatly decreased grooming behavior (of themselves and of others); greatly reduced affective interactions with others, regardless of whether they are males, females or infants; diminished facial expressions and vocalizations; impaired maternal behavior; and sexual indifference. While they can move normally, they fail to relate to the other animals in the troop to which they belonged before the operation, and the other animals fail to relate to them. The other animals can, however, relate normally to monkeys that develop major physical defects such as paralysis but that do not have prefrontal damage. Although the paralytic monkeys seem more disabled than the monkeys with prefrontal damage, they seek and receive the support of their peers.

It is fair to assume that monkeys with prefrontal damage can no longer follow the complex social conventions characteristic of the organization of a monkey troop (hierarchical relations of its different members, dominance of certain females and males over other members, and so on[22]). It is likely that they fail in terms of "social cognition" and in terms of "social behavior" and that the other animals respond in kind. Remarkably, monkeys with damage in motor cortex, but not in prefrontal cortex, have no such difficulties.

Monkeys with bilateral ablations of the anterior sector of the temporal lobe (from operations that do *not* damage the amygdala) reveal some impairment of social behavior, but to a far lesser degree than monkeys with prefrontal damage. In spite of the marked neurobiological differences between monkey and chimpanzee, and between chimpanzee and human, there is a shared essence to the defect caused by prefrontal damage: Personal and social behavior is severely compromised.[23]

The work of Fulton and Jacobsen provides other important evidence. As was mentioned, the aim of their studies was to understand learning and memory, and from that standpoint their results constitute a landmark. The purpose of one task the researchers set for the chimpanzees was the learning of an association between a rewarding stimulus and the position of that stimulus in space. Their classic experiment went like this: One animal had before her, within arm's reach, two wells. A desirable piece of food was placed in one of the wells, in full view of the animal, and then both wells were covered so that the food was no longer visible. After a delay of several seconds, the animal had to reach the well in which the food was hidden and avoid the empty one. The normal animal held the knowledge of where the food was for the entire duration of the delay and then made the appropriate move to obtain the food. But after prefrontal damage, the animals could no longer perform the task. As soon as the stimulus was out of sight, it seems it was also out of mind. These findings became the cornerstone for the subsequent neurophysiologic explorations of prefrontal cortex by Patricia Goldman-Rakic and Joaquim Fuster.[24]

A recent and especially relevant finding for my argument concerns the concentration of one of the chemical receptors for serotonin in the ventromedial sector of the prefrontal cortex and in the amygdala. Serotonin is one of the main neurotransmitters, substances whose actions contribute to virtually all aspects of behavior and cognition (other key neurotransmitters are dopamine, norepinephrine, and acetylcholine; they are all delivered from neurons located in small nuclei of the brain stem or the basal forebrain, whose axons terminate in the neocortex, the cortical and subcortical components of the limbic system, the basal ganglia, and the thalamus). One of the roles of serotonin in primates is the inhibition of aggressive behavior (curiously, it has other roles in other species). In experimental animals, when the neurons in which serotonin originates are blocked from delivering it, one consequence is that the animals behave impulsively and aggressively. In general, enhancing serotonin function reduces aggression and favors social behavior.

In this context it is important to note, as shown in the work of Michael Raleigh,[25] that in monkeys whose behavior is socially well tuned (as measured by displays of cooperation, grooming, and proximity to others), the number of serotonin-2 receptors is extremely high in the ventromedial frontal lobe, the amygdala, and the medial temporal cortices in its vicinity, but not elsewhere in the brain; and that in monkeys exhibiting noncooperative, antagonistic behavior, the opposite is true. This finding reinforces the system connection between ventromedial prefrontal cortices and amygdala that I have suggested on the basis of neuropsychological results, and it relates these regions to social behavior, the principal domain affected in my patients' flawed decision-making. (The reason why the serotonin receptors identified in this study are marked as "serotonin-2" is because there are many different types of serotonin receptor, no less than 14 in fact.)

An Aside on Neurochemical Explanations

When it comes to explaining behavior and mind, it is not enough to mention neurochemistry. We must know whereabouts the chemistry is, in the system presumed to cause a given behavior. Without knowing the cortical regions or nuclei where the chemical acts within the system, we have no chance of ever understanding how it modifies the system's performance (and keep in mind that such understanding is only the first step, prior to the eventual elucidation of how more fine-grained circuits operate). Moreover, the neural explanation only begins to be useful when it addresses the *results* of the operation of a given system on yet another system. The important finding described above should not be demeaned by superficial statements to the effect that serotonin alone "causes" adaptive social behavior and its lack "causes" aggression. The presence or absence of serotonin in specific brain systems having specific serotonin receptors does change their operation; and such change, in turn, modifies the operation of yet other systems, the result of which will ultimately be expressed in behavioral and cognitive terms.

These comments about serotonin are especially pertinent, given the recent high visibility of this neurotransmitter. The popular antidepressant Prozac, which acts by blocking the reuptake of serotonin and probably increasing its availability, has received wide attention; the notion that low serotonin levels might be correlated with a tendency toward violence has surfaced in the popular press. The problem is that it is not the absence or low amount of serotonin per se that "causes" a certain manifestation. Serotonin is part of an exceedingly complicated mechanism which operates at the level of molecules, synapses, local circuits, and systems, and in which sociocultural factors, past and present, also intervene powerfully. A satisfactory explanation can arise only from a more comprehensive view of the entire process, in which the relevant variables of a specific problem, such as depression or social adaptability, are analyzed in detail.

On a practical note: The solution to the problem of social violence will not come from addressing only social factors and ignoring neurochemical correlates, nor will it come from blaming one neurochemical correlate alone. Consideration of *both* social and neurochemical factors is required, in appropriate measure.

CONCLUSION

The human evidence discussed in this section suggests a close bond between a collection of brain regions and the processes of reasoning and decision making. Animal studies have revealed some of the same bonds involving some of the same regions. By combining evidence from both human and animal studies we can now itemize a few facts about the roles of the neural systems we have identified.

First, these systems are certainly involved in the processes of reason in the broad sense of the term. Specifically, they are involved in planning and deciding.

Second, a subset of these systems is associated with planning and deciding behaviors that one might subsume under the rubric "personal and social." There is a hint that these systems are related to the aspect of reason usually designated as rationality.

Third, the systems we have identified play an important role in the processing of emotions.

Fourth, the systems are needed to hold in mind, over an extended period of time, the image of a relevant but no longer present object.

Why should such disparate roles come together in a circum-scribed sector of the brain? What can possibly be shared by planning and making personal and social decisions; processing emotion; and holding an image in mind, in the absence of the thing it represents?

Part

2

Five

Assembling an Explanation

A MYSTERIOUS ALLIANCE

THE INVESTIGATION OF patients with newly acquired impairments of reasoning and decision making described in part I led to the identification of a specific set of brain systems that were consistently damaged in those patients. It also identified an apparently odd collection of neuropsychological processes that depended on the integrity of those systems. What connects those processes among themselves in the first place, and what links them to the neural systems outlined in the previous chapter? The following paragraphs offer some provisional answers.

First, reaching a decision about the typical personal problem posed in a social environment, which is complex and whose outcome is uncertain, requires both broad-based knowledge and reasoning strategies to operate over such knowledge. The broad knowledge includes facts about objects, persons, and situations in the external world. But because personal and social decisions are inextricable from survival, the knowledge also includes facts and mechanisms concerning the

...r the organism as a whole. The reasoning strategies ...und goals, options for action, predictions of future out- ...d plans for implementation of goals at varied time scales. ...nd, the processes of emotion and feeling are part and parcel of the neural machinery for biological regulation, whose core is constituted by homeostatic controls, drives, and instincts.

Third, because of the brain's design, the requisite broad-based knowledge depends on numerous systems located in relatively separate brain regions rather than in one region. A large part of such knowledge is recalled in the form of images at many brain sites rather than at a single site. Although we have the illusion that everything comes together in a single anatomical theater, recent evidence suggests that it does not. Probably the relative simultaneity of activity at different sites binds the separate parts of the mind together.

Fourth, since knowledge can be retrieved only in distributed, parcellated manner, from sites in many parallel systems, the operation of reasoning strategies requires that the representation of myriad facts be held active in a broad parallel display for an extended period of time (in the very least for several seconds). In other words, the images over which we reason (images of specific objects, actions, and relational schemas; of words which help translate the latter into language form) not only must be "in focus"—something achieved by attention—but also must be "held active in mind"—something achieved by high-order working memory.

I suspect that the mysterious alliance of the processes uncovered at the end of the previous chapter is due in part to the nature of the problem the organism is attempting to solve, and in part to the brain's design. Personal and social decisions are fraught with uncertainty and have an impact on survival, directly or indirectly. Thus they require a vast repertoire of knowledge concerning the external world and the world within the organism. However, since the brain holds and retrieves knowledge in spatially segregated rather than integrated manner, they also require attention and working memory so that the component of knowledge that is retrieved as a display of images can be manipulated in time.

As for why the neural systems we identified overlap so blatantly, I suspect evolutionary convenience is the answer. If basic biological regulation is essential to the guidance of personal and social behavior, then a brain design likely to have prevailed in natural selection may have been one in which the subsystems responsible for reasoning and decision making would have remained intimately interlocked with those concerned with biological regulation, given their shared involvement in the business of survival.

The general explanation previewed in these answers is a first approximation to the questions posed by Phineas Gage's case. What in the brain allows humans to behave rationally? How does it work? I usually resist subsuming the effort to answer these questions with the expression "neurobiology of rationality," because it sounds official and pretentious, but that is it, in a nutshell: the beginnings of a neurobiology of human rationality at the level of large-scale brain systems.

My plan in this second part of the book is to address the plausibility of the general explanation outlined above and present a testable hypothesis derived from it. Because of the wide ramifications of the subject, however, I restrict the discussion to a select number of matters that I regard as indispensable to make the ideas intelligible.

This chapter is a bridge between the facts of part I and the interpretations I offer later. The traversal—I hope you don't come to regard it as an interruption—has several purposes: to survey notions to which I will appeal frequently (e.g., organism, body, brain, behavior, mind, state); to discuss briefly the neural basis of knowledge with an emphasis on its parcellated nature and its dependence on images; and to make comments on neural development. I will not be exhaustive (for instance, a discussion on learning or on language would have been appropriate and useful, but neither topic is indispensable for the aim I have in mind); I will not offer a textbook treatment of any topic; and I will not justify every opinion I express. Remember, this is a conversation.

Subsequent chapters return to our main story and will address biological regulation, its expression in emotion and feeling, and the mechanisms whereby emotion and feeling may be used in decision making.

Before going any further, I must repeat something I said in the introduction. The text is an open-ended exploration rather than a catalogue of agreed-upon facts. I am considering hypotheses and empirical tests, not making affirmations of certainty.

OF ORGANISMS, BODIES, AND BRAINS

Whatever questions one may have about who we are and why we are as we are, it is certain that we are complex living organisms with a body proper ("body" for short) and a nervous system ("brain" for short). Whenever I refer to the body I mean the organism minus the neural tissue (the central and peripheral components of the nervous system), although in the conventional sense the brain is also part of the body.

The organism has a structure and myriad components. It has a bony skeleton with many parts, connected by joints and moved by muscles; it has numerous organs combined in systems; it has a boundary or membrane marking its outer limit, made largely of skin. On occasion I will refer to organs—blood vessels, organs in the head, chest and abdomen, the skin—as "viscera" (singular "viscus"). Again, in the conventional sense, the brain would be included, but I exclude it here.

Each part of the organism is made of biological tissues, which are in turn made of cells. Each cell is made of numerous molecules arranged to create a skeleton for the cell (cytoskeleton), numerous organs and systems (cell nuclei and varied organelles), and an overall boundary (cell membrane). The complexity of structure and function is daunting when we look at one of those cells in operation, and staggering when we look at an organ system in the body.

STATES OF ORGANISMS

In the discussion ahead there are many references to "body states" and "mind states." Living organisms are changing continuously, assuming a succession of "states," each defined by varied patterns of ongoing activity in all of its components. You might picture this as a composite of the actions of a slew of people and objects operating within a circumscribed area. Imagine yourself in a large airport terminal, looking around, inside and outside. You see and hear the constant bustle from many different systems: people boarding or leaving aircraft, or just sitting or standing; people strolling or walking by with seeming purpose; planes taxiing, taking off, landing; mechanics and baggage handlers going about their business. Now imagine that you freeze the frame of this ongoing video or that you take a wide-angle snapshot of the entire scene. What you get in the frozen frame or in the still snapshot is the image of a *state*, an artificial, momentary slice of life, indicating what was going on in the various organs of a vast organism during the time window defined by the camera's shutter speed. (In reality, things are a bit more complicated than this. Depending on the scale of analysis, the states of organisms may be discrete units or merge continuously.)

BODY AND BRAIN INTERACT: THE ORGANISM WITHIN

The brain and the body are indissociably integrated by mutually targeted biochemical and neural circuits. There are two principal routes of interconnection. The route usually thought of first is made of sensory and motor peripheral nerves which carry signals from every part of the body to the brain, and from the brain to every part of the body. The other route, which comes less easily to mind although it is far older in evolution, is the bloodstream; it carries chemical signals such as hormones, neurotransmitters, and modulators.

Even a simplified summary reveals the intricacy of the relationships:

1. Nearly every part of the body, every muscle, joint, and internal organ, can send signals to the brain via the peripheral nerves. Those signals enter the brain at the level of the spinal cord or the brain stem, and eventually are carried inside the brain, from neural station to neural station, to the somatosensory cortices in the parietal lobe and insular regions.

2. Chemical substances arising from body activity can reach the brain via the bloodstream and influence the brain's operation either directly or by activating special brain sites such as the subfornical organ.

3. In the opposite direction, the brain can act, through nerves, on all parts of the body. The agents for those actions are the autonomic (or visceral) nervous system and the musculoskeletal (or voluntary) nervous system. The signals for the autonomic nervous system arise in the evolutionarily older regions (the amygdala, the cingulate, the hypothalamus, and the brain stem), while the signals for the musculoskeletal system arise in several motor cortices and subcortical motor nuclei, of different evolutionary ages.

4. The brain also acts on the body by manufacturing or ordering the manufacture of chemical substances released in the bloodstream, among them hormones, transmitters, and modulators. I will say more about these in the next chapter.

When I say that body and brain form an indissociable organism, I am not exaggerating. In fact, I am oversimplifying. Consider that the brain receives signals not only from the body but, in some of its sectors, from parts of itself that receive signals from the body! The organism constituted by the brain-body partnership interacts with the environment as an ensemble, the interaction being of neither the body nor the brain alone. But complex organisms such as ours do more than just interact, more than merely generate the spontaneous or reactive external responses known collectively as behavior. They also generate internal responses, some of which constitute images (visual, auditory, somatosensory, and so on), which I postulate as the basis for mind.

OF BEHAVIOR AND MIND

Many simple organisms, even those with only a single cell and no brain, perform actions spontaneously or in response to stimuli in the environment; that is, they produce behavior. Some of these actions are contained in the organisms themselves, and can be either hidden to observers (for instance, a contraction in an interior organ), or externally observable (a twitch, or the extension of a limb). Other actions (crawling, walking, holding an object) are directed at the environment. But in some simple organisms and in all complex organisms, actions, whether spontaneous or reactive, are caused by commands from a brain. (Organisms with a body and no brain, but capable of movement, it should be noted, preceded and then coexisted with organisms that have both body and brain.)

Not all actions commanded by a brain are caused by deliberation. On the contrary, it is a fair assumption that most so-called brain-caused actions being taken at this very moment in the world are not deliberated at all. They are simple responses of which a reflex is an example: a stimulus conveyed by one neuron leading another neuron to act.

As organisms acquired greater complexity, "brain-caused" actions required more intermediate processing. Other neurons were interpolated between the stimulus neuron and the response neuron, and varied parallel circuits were thus set up, but it did not follow that the organism with that more complicated brain necessarily had a mind. Brains can have many intervening steps in the circuits mediating between stimulus and response, and still have no mind, if they do not meet an essential condition: the ability to display images internally and to order those images in a process called thought. (The images are not solely visual; there are also "sound images," "olfactory images," and so on.) My statement about behaving organisms can now be completed by saying that not all have minds, that is, not all have mental phenomena (which is the same as saying that not all have cognition or cognitive processes). Some organisms have both behavior and cognition. Some have intelligent actions but no mind. No organism seems to have mind but no action.

My view then is that having a mind means that an organism forms neural representations which can become images, be manipulated in a process called thought, and eventually influence behavior by helping predict the future, plan accordingly, and choose the next action. Herein lies the center of neurobiology as I see it: the process whereby neural representations, which consist of biological modifications created by learning in a neuron circuit, become images in our minds; the process that allows for invisible microstructural changes in neuron circuits (in cell bodies, dendrites and axons, and synapses) to become a neural representation, which in turn becomes an image we each experience as belonging to us.

To a first approximation, the overall function of the brain is to be well informed about what goes on in the rest of the body, the body proper; about what goes on in itself; and about the environment surrounding the organism, so that suitable, survivable accommodations can be achieved between organism and environment. From an evolutionary perspective, it is not the other way around. If there had been no body, there would have been no brain. Incidentally, the simple organisms with just body and behavior but no brain or mind are still here, and are in fact far more numerous than humans by several orders of magnitude. Think of the many happy bacteria such as *Escherichia coli* now living inside each of us.

ORGANISM AND ENVIRONMENT INTERACT: TAKING ON THE WORLD WITHOUT

If body and brain interact with each other intensely, the organism they form interacts with its surroundings no less so. Their relations are mediated by the organism's movement and its sensory devices.

The environment makes its mark on the organism in a variety of ways. One is by stimulating neural activity in the eye (inside which is the retina), the ear (inside which are the cochlea, a sound-sensing device, and the vestibule, a balance-sensing device), and the myriad nerve terminals in the skin, taste buds, and nasal mucosa. Nerve terminals send signals to circumscribed entry points in the brain, the so-called early sensory cortices of vision, hearing, somatic sensations, taste, and olfaction. Picture them as a sort of safe harbor where signals can arrive. Each early sensory region (early visual cortices, early auditory cortices, and so forth) is a collection of several areas, and there is heavy cross-signaling among the aggregate of areas in each early sensory collection, as you can see in Fig. 5-1. Later in this chapter I will suggest that these closely interlocked sectors are the basis for topographically organized representations, the source of mental images.

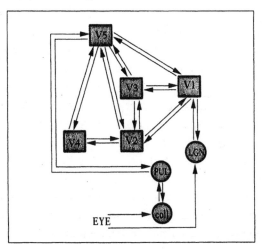

Figure 5-1. A simplified diagram of some interconnections among the "early visual cortices" (V1, V2, V3, V4, V5) and three visually related subcortical structures: lateral geniculate nucleus (LGN); the pulvinar (PUL) and the superior colliculus (coll). V1 is also known as the "primary" visual cortex, and corresponds to Brodmann's area 17. Note that most of the components in this system are interconnected by feedforward and feedback neuron projections (arrowed lines). The visual input to the system comes from the eye via the LGN and colliculus. The outputs of this system arise from many of the components, in parallel (e.g., from V4, V5, and so on), toward cortical as well as subcortical targets.

In turn, the organism acts on the environment by means of movements in the whole body, the limbs, and the vocal apparatus which are controlled by the M_1, M_2, and M_3 cortices (the cortices in which body-aimed movements also arise), with the help of several subcortical motor nuclei. There are, then, brain sectors where signals from the body proper or the body's sense organs arrive continuously. These "input" sectors are anatomically separate and do not communicate with one another directly. There are also brain sectors where motor and chemical signals arise; among these "output" sectors are the brain stem and hypothalamic nuclei, and the motor cortices.

An Aside on the Architecture of Neural Systems

Pretend you are designing the human brain from scratch and have penciled in all the harbors to which you would ferry the many sensory signals. Would you not want to merge the signals from different sensory sources, say, vision and hearing, as rapidly as possible so that the brain could generate "integrated representations" of things simultaneously seen and heard? Would you not want to connect those representations to motor controls so that the brain could respond effectively to them? I assume your answer is a resounding yes, but that has not been nature's answer. As a landmark study of neuronal connections by E. G. Jones and T. P. S. Powell showed, some two decades ago, nature does not let the sensory harbors talk to each other *directly*, and it does not permit them to talk to motor controls *directly* either.[1] At the level of the cerebral cortex, for instance, each collection of early sensory areas must talk first to a variety of interposed regions, which talk to regions farther away, and so forth. The talking is carried out by forward-projecting axons, or feedforward projections, which converge to regions downstream, which themselves converge to other regions.

It may seem that these multiple, parallel, converging streams terminate at some apex points, such as the cortex nearest to the hippocampus (the entorhinal cortex), or some sectors of the prefrontal cortex (the dorsolateral or ventromedial). But this is not quite accurate. For one thing, they never "terminate" as such, because, from the vicinity of each point to which they project forward, there is a reciprocal projection backward. It is appropriate to say that signals in the stream move both forward *and* backward. Instead of a forward-moving stream, one finds loops of feedforward and feedback projections, which can create a perpetual recurrence.

Another reason why the streams do not "terminate" in the proper sense is that out of some of their stations, especially those that are forward placed, there are direct projections to motor controls.

Thus communication among input sectors and between input and output sectors is not direct but intermediate, and it uses a complex architecture of interconnected neuron assemblies. At the level of the cerebral cortex those assemblies are cortical regions located within varied association cortices. But intermediate communication occurs also via large subcortical nuclei such as those in the thalamus and basal ganglia, and via small nuclei such as those in the brain stem.

In short, the number of brain structures located between the input and output sectors is quite large, and the complexity of their connection patterns immense. The natural question is: What happens in all those "interposed" structures, what does all that complexity buy us? The answer is that activity there, together with that of the input and output areas, momentarily constructs and stealthily manipulates the images in our minds. On the basis of those images, about which I will say more in the pages ahead, we can interpret the signals brought in at the early sensory cortices so that we can organize them as concepts and categorize them. We can acquire strategies for reasoning and decision making; and we can select a motor response from the menu available in our brain, or formulate a new motor response, a willed, deliberated composition of actions, which can range from pounding on a table, to hugging a child, to writing a letter to the editor, or to playing Mozart on the piano.

In between the brain's five main sensory input sectors and three main output sectors lie the association cortices, the basal ganglia, the thalamus, the limbic system cortices and limbic nuclei, and the brain stem and cerebellum. Together, this "organ" of information and government, this great collection of systems, holds both innate and acquired knowledge about the body proper, the outside world, and the brain itself as it interacts with body proper and outside world. This knowledge is used to deploy and manipulate motor outputs and mental outputs, the images that constitute our thoughts. I believe that this repository of facts and strategies for their manipulation is stored, dormantly and abeyantly, in the form of "dispositional representations" ("dispositions," for short) in the in-between brain sectors. Biological regulation, memory of previous states, and planning of future actions result from cooperative activity not just in early sensory and motor cortices but also in the in-between sectors.

AN INTEGRATED MIND FROM PARCELLATED ACTIVITY

One common false intuition shared by many who enjoy thinking about how the brain works is that the many strands of sensory processing experienced in the mind—sights and sounds, taste and aroma, surface texture and shape—all "happen" in a single brain structure. Somehow it stands to reason that what is together in the mind is together at one place in the brain where different sensory aspects mingle. The usual metaphor has something to do with a large CinemaScope screen equipped for glorious Technicolor projection, stereophonic sound, and perhaps a track for smell too. Daniel Dennett has written extensively about this concept which he dubbed "Cartesian theater," and has argued persuasively, on cognitive grounds, that the Cartesian theater cannot exist.[2] I too, on neuroscientific grounds, maintain that it is a false intuition.

I will summarize here my reasons, which I have discussed elsewhere at length.[3] My main argument against the idea of an integrative brain site is that there is no single region in the human brain equipped to process, simultaneously, representations from all the sensory modalities active when we experience simultaneously, say, sound, movement, shape, and color in perfect temporal and spatial registration.

We are beginning to glean where the construction of images for each separate modality is likely to take place, but nowhere can we find a single area toward which all of those separate products would be projected in exact registration.

It is true that there are a few brain regions where signals from many different early sensory regions can converge. A few of those convergence regions actually receive a wide variety of polymodal signals, for instance, the entorhinal and perirhinal cortices. But the kind of integration those regions can produce using such signals is unlikely to be the one that forms the base for the integrated mind. For one thing, damage to those higher-order convergence regions, even when it occurs in both hemispheres, does not preclude "mind" integration at all, although it causes other detectable neuro-psychological consequences such as learning impairments.

It is perhaps more fruitful to think that our strong sense of mind integration is created from the concerted action of large-scale systems by synchronizing sets of neural activity in separate brain regions, in effect a trick of timing. If activity occurs in anatomically separate brain regions, but if it does so within approximately the same window of time, it is still possible to link the parts behind the scenes, as it were, and create the impression that it all happens in the same place. Note that this is in no way an explanation of how time does binding, but rather a suggestion that timing is an important part of the mechanism. The idea of integration by time has surfaced over the past decade and now appears prominently in the work of a number of theorists.[4]

If the brain does integrate separate processes into meaningful combinations by means of time, this is a sensible and economical solution but not one without risks and problems. The main risk is mistiming. Any malfunction of the timing mechanism would be likely to create spurious integration or *dis*integration. This may be indeed what happens in states of confusion caused by head injury, or in some symptoms of schizophrenia and other diseases. The fundamental problem created by time binding has to do with the requirement for maintaining focused activity at different sites for as long as necessary for meaningful combinations to be made and for reasoning and decision making to take place. In other words, time binding requires powerful and effective mechanisms of attention and working memory, and nature seems to have agreed to provide them.

Each sensory system appears equipped to provide its own local attention and working-memory devices. But when it comes to the processes of global attention and working memory, human studies as well as animal experiments suggest that the prefrontal cortices and some limbic system structures (the anterior cingulate) are essential.[5] The mysterious connection between the processes and brain systems discussed at the beginning of this chapter may be clearer now.

IMAGES OF NOW, IMAGES OF THE PAST, AND IMAGES OF THE FUTURE

The factual knowledge required for reasoning and decision making comes to the mind in the form of images. Let us look, however briefly, at the possible neural substrate of those images.

If you look out the window at the autumn landscape, or listen to the music playing in the background, or run your fingers over a smooth metal surface, or read these words, line after line down this page, you are perceiving, and thereby forming images of varied sensory modalities. The images so formed are called *perceptual images*.

But you may stop attending to that landscape or music or surface or text, distract yourself from it, and turn your thoughts elsewhere. Perhaps you are now thinking of your Aunt Maggie, or the Eiffel Tower, or the voice of Plácido Domingo, or of what I just said about images. Any of those thoughts is also constituted by images, regardless of whether they are made up mostly of shapes, colors, movements, tones, or spoken or unspoken words. Those images, which occur as you conjure up a remembrance of things past, are known as *recalled images,* so as to distinguish them from the perceptual variety.

By using recalled images you can bring back a particular type of past image, one formed when you planned something that has not yet happened but that you intend to have happen, for example, reorganizing your library come this weekend. As the planning process unfolded, you were forming images of objects and movements, and consolidating a memory of that fiction in your mind. Images of something that has not yet happened and that may in fact never come to pass are no different in nature from the images you hold of something that already has happened. They constitute the memory of a possible future rather than of the past that was.

These various images—perceptual, recalled from real past, and recalled from plans of the future—are constructions of your organism's brain. All that you can know for certain is that they are real to your self, and that other beings make comparable images. We share our image-based concept of the world with other humans, and even with some animals; there is a remarkable consistency in the constructions different individuals make of the essential aspects of the environment (textures, sounds, shapes, colors, space). If our organisms were designed differently, the constructions we make of the world around us would be different as well. We do not know, and it is improbable that we will ever know, what "absolute" reality is like.

How do we come to create these marvelous constructions? It appears they are concocted by a complex neural machinery of perception, memory, and reasoning. Sometimes the construction is paced from the world outside the brain, that is, from the world inside our body or around it, with a bit of help from past memory. That is the case when we generate perceptual images. Sometimes the construction is directed entirely from within our brain, by our sweet and silent thought process, from the top down, as it were. That is the case, for instance, when we recall a favorite melody, or recall visual scenes with our eyes closed and covered, whether the scenes are a replaying of a real event or an imagined one.

But the neural activity that is most closely related to the images we experience occurs in early sensory cortices and not in the other regions. The activity in the early sensory cortices, whether it is engaged by perception or by recall of memories, is a result, so to speak, of complex processes operating behind the scenes, in numerous regions of the cerebral cortex and of neuron nuclei beneath the cortex, in basal ganglia, brain stem, and elsewhere. In short: *Images are based directly on those neural representations, and only those, which are organized topographically and which occur in early sensory cortices.* But they are formed either under the control of sensory receptors oriented to the brain's outside (e.g., a retina), or under the control of dispositional representations (dispositions) contained inside the brain, in cortical regions and subcortical nuclei.

Forming Perceptual Images

How are images formed when you are perceiving something in the world, a landscape, say, or in the body, for instance, a pain in your right elbow? In both cases, there is a first step which is necessary but not sufficient: Signals from the appropriate body sector (eye and retina, in one case; nerve terminals in the elbow joint, in the other) are carried by neurons, down their axons and across several electrochemical synapses, into the brain. The signals are delivered to the early sensory cortices.* For signals from the retina this will happen in the early visual cortices, located at the back of the brain in the occipital lobe. For signals from the elbow joint, this will happen in the early somatosensory cortices in the parietal and insular regions, part of the brain sector that is damaged in

---

*The workings of the perceptual machinery within those early cortices are beginning to be understood. Studies of the visual system, for which a large quantity of neuroanatomical, neurophysiological, and psychophysical data have now been gathered, lead the way, but there is a wealth of new findings in somatosensory and auditory systems. These cortices form a dynamic coalition, and the topographically organized representations they generate change with the type and amount of input, as the work of several researchers has demonstrated.[6]

anosognosia. Note again that this is a *collection* of areas rather than one center. The areas that are part of the collection are individually complex and the mesh of interconnections they form is even more so. The topographically organized representations result from the concerted interaction of these areas, not from one of them only. There is nothing phrenological about this idea.

When all or most early sensory cortices of a given sensory modality are destroyed, the ability to form images in that modality vanishes. Patients deprived of early visual cortices are not able to see much. (Some residual sensory capacities are preserved in those patients, probably because cortical and subcortical structures related to the sensory modality are intact. After extensive destruction of the early visual cortices, some patients can point to light targets that they profess not to see; they have what is known as blindsight. The parietal cortices, the superior colliculi, and the thalamus are just a few of the structures presumably involved in these processes.) The perceptual defect can be quite specific. After damage to one of the subsystems within the early visual cortices, for example, there may be a loss of the ability to perceive color; this loss may be complete, or an attenuation, such that patients perceive colors as drained out. Affected patients see shape, movement, and depth, but not color. In this condition, achromatopsia, patients construct the universe in shades of gray.

Although the early sensory cortices and the topographically organized representations they form are necessary for images to occur in consciousness, they do not, however, appear to be sufficient. In other words, if our brains would simply generate fine topographically organized representations and do nothing else with those representations, I doubt we would ever be conscious of them as images. How would we know they are *our* images? Subjectivity, a key feature of consciousness, would be missing from such a design. Other conditions must be met.

In essence those neural representations must be correlated with those which, moment by moment, constitute the neural basis for the self. This issue will surface again in chapters 7 and 10, but let me say at this point that the self is not the infamous homunculus, a little person inside our brain perceiving and thinking about the images the brain forms. It is, rather, a perpetually re-created neurobiological state. Years of justified attack on the homunculus concept have made many theorists equally fearful of the concept of self. But the neural self need not be homuncular at all. What should cause some fear, actually, is the idea of a selfless cognition.

STORING IMAGES AND FORMING IMAGES IN RECALL

Images are *not* stored as facsimile pictures of things, or events, or words, or sentences. The brain does not file Polaroid pictures of people, objects, landscapes; nor does it store audiotapes of music and speech; it does not store films of scenes in our lives; nor does it hold the type of cue cards and TelePrompTer transparencies that help politicians earn their daily bread. In brief, there seem to be no permanently held pictures of anything, even miniaturized, no microfiches or microfilms, no hard copies. Given the huge amount of knowledge we acquire in a lifetime, any kind of facsimile storage would probably pose insurmountable problems of capacity. If the brain were like a conventional library, we would run out of shelves just as conventional libraries do. Furthermore, facsimile storage also poses difficult problems of retrieval efficiency. We all have direct evidence that whenever we recall a given object, or face, or scene, we do not get an exact reproduction but rather an *interpretation*, a newly reconstructed version of the original. In addition, as our age and experience change, versions of the same thing evolve. None of this is compatible with rigid, facsimile representation, as the British psychologist Frederic Bartlett noted several decades ago, when he first proposed that memory is essentially reconstructive.[7]

Yet the denial that permanent pictures of anything can exist in the brain must be reconciled with the sensation, which we all share, that we *can* conjure up, in our mind's eye or ear, approximations of images we previously experienced. That these approximations are not accurate, or are less vivid than the images they are meant to reproduce, does not contradict this fact.

A tentative answer to this problem suggests that these mental images are momentary constructions, *attempts at replication* of patterns that were once experienced, in which the probability of exact replication is low but the probability of substantial replication can be higher or lower, depending on the circumstances in which the images were learned and are being recalled. These recalled images tend to be held in consciousness only fleetingly, and although they may appear to be good replicas, they are often inaccurate or incomplete. I suspect that explicit recalled mental images arise from the transient synchronous activation of neural firing patterns largely in the same early sensory cortices where the firing patterns corresponding to perceptual representations once occurred. The activation results in a topographically organized representation.

There are several arguments in favor of this notion, and some evidence. In the condition known as achromatopsia, described above, local damage in the early visual cortices causes not only loss of color perception but also loss of color imagery. If you are achromatopsic, you can no longer *imagine* color in your mind. If I ask you to imagine a banana, you will be able to picture its shape but not its color; you will see it in shades of gray. If "color knowledge" were stored elsewhere, in a system separate from the one that supports "color perception," achromatopsic patients would imagine color even when they cannot perceive it in an external object. But they do not.

Patients with extensive damage to the early visual cortices lose their ability to generate visual imagery. Yet they can still recall knowledge about tactile and spatial properties of objects, and they can still recall sound images.

Preliminary studies of visual recall using positron emission tomography (PET), a neuroimaging technique, and functional magnetic resonance (FMR) support this idea. Steven Kosslyn and his group, and Hanna Damasio, Thomas Grabowski, and their colleagues, have found that recollection of visual images activates the early visual cortices, among other areas.[8]

. . .

How do we form the topographically organized representations needed to experience recalled images? I believe those representations are constructed momentarily under the command of acquired *dispositional* neural patterns elsewhere in the brain. I use this term because what they do, quite literally, is order other neural patterns about, make neural activity happen elsewhere, in circuits that are part of the same system and with which there is a strong neuronal interconnection. Dispositional representations exist as potential patterns of neuron activity in small ensembles of neurons I call "convergence zones"; that is, they consist of a set of neuron firing dispositions within the ensemble. The dispositions related to recall-able images were acquired through learning, and thus we can say they constitute a memory. The convergence zones whose dispositional representations can result in images when they fire back to early sensory cortices are located throughout the higher-order association cortices (in occipital, temporal, parietal, and frontal regions), and in basal ganglia and limbic structures.

What dispositional representations hold in store in their little commune of synapses is not a picture per se, but a means to reconstitute "a picture." If you have a dispositional representation for the face of Aunt Maggie, that representation contains not her face as such, but rather the firing patterns which trigger the momentary reconstruction of an approximate representation of Aunt Maggie's face, in early visual cortices.

The several dispositional representations that would need to fire back, more or less synchronously, for Aunt Maggie's face to show up in the scopes of your mind, are located in several visual and higher-order association cortices (mostly, I suspect, in occipital and temporal regions).[9] The same arrangement would apply in the auditory realm. There are dispositional representations for Aunt Maggie's voice in auditory association cortices, which can fire back to early auditory cortices and generate momentarily the approximate representation of Aunt Maggie's voice.

There is not just one hidden formula for this reconstruction. Aunt Maggie as a complete person does not exist in one single site of your brain. She is distributed all over it, in the form of many dispositional representations, for this and that. And when you conjure up remembrances of things Maggie, and she surfaces in various early cortices (visual, auditory, and so on) in topographic representation, she is still present only in separate views during the time window in which you construct *some* meaning of her person.

Were you to fall inside somebody's *visual dispositional* representations for Aunt Maggie, in an imaginary experiment fifty years from now, I predict you would see nothing resembling Aunt Maggie's face, because dispositional representations are *not* topographically organized. But if you were to inspect the patterns of activity occurring in that somebody's early visual cortices, within about a hundred milliseconds after the convergence zones for Aunt Maggie's face fired back, you probably would be able to see patterns of activity that had some relation to the geography of Aunt Maggie's face. There would be *consistency* between what you knew of her face, and the pattern of activity you would find in the early visual cortical circuitry of somebody who knew her too and was thinking of her.

There is already evidence suggesting that this would be so. Using a neuroanatomical imaging method, R. B. H. Tootell has shown that when a monkey sees certain shapes, such as a cross or square, the activity of neurons in early visual cortices will be topographically organized in a pattern that conforms to the shapes the monkey is viewing.[10] In other words, an independent observer looking at the external stimulus and at the pattern of brain activity recognizes structural similarity. (See Fig. 5-2.) Similar reasoning can be applied to Michael Merzenich's findings about the dynamic patterns of body representation in the somatosensory cortices.[11] Note, however, that having such a representation in the cerebral cortex is *not* equivalent to being conscious of it, as I pointed out earlier. It is necessary but not sufficient.

What I am calling a dispositional representation is a dormant firing potentiality which comes to life when neurons fire, with a

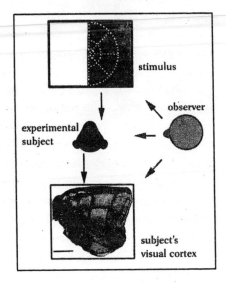

Figure 5-2. An observer looking at the stimulus presented to an experimental animal, who subsequently would look at the activation caused by that stimulus in the animal's visual cortex, would discover a remarkable consistency between the shape of the stimulus and the shape of the neural activity pattern in one of the layers of the primary visual cortex (layer 4C). The stimulus and brain image came from the work of Roger Tootell who performed this experiment.

particular pattern, at certain rates, for a certain amount of time, and toward a particular target which happens to be another ensemble of neurons. Nobody knows what the "codes" contained in the ensemble might look like, despite the many new findings that have been amassed in the study of synaptic modification. But this much appears likely: The firing patterns result from the strengthening or weakening of synapses, and that, in turn, results from functional changes occurring at microscopic level within the fiber branches of neurons (axons and dendrites).[12]

Dispositional representations exist in potential state, subject to activation, like the town of Brigadoon.

KNOWLEDGE IS EMBODIED IN DISPOSITIONAL REPRESENTATIONS

Dispositional representations constitute our full repository of knowledge, encompassing both innate knowledge and knowledge acquired by experience. Innate knowledge is based on dispositional representations in hypothalamus, brain stem, and limbic system. You can conceptualize it as commands about biological regulation which are required for survival (e.g., the control of metabolism, drives, and instincts). They control numerous processes, but by and large they do not become images in the mind. These will be discussed in the next chapter.

Acquired knowledge is based on dispositional representations in higher-order cortices and throughout many gray-matter nuclei beneath the level of the cortex. Some of those dispositional representations contain records for the imageable knowledge that we can recall and which is used for movement, reason, planning, creativity; and some contain records of rules and strategies with which we operate on those images. The acquisition of new knowledge is achieved by continuous modification of such dispositional representations.

When dispositional representations are activated, they can have various results. They can fire other dispositional representations to which they are strongly related by circuit design (dispositional representations in the temporal cortex, for example, could fire dispositional representations in the occipital cortex which are part of the same *strengthened* systems). Or they can generate a topographically organized representation, by firing back to early sensory cortices directly, or by activating other dispositional representations in the same *strengthened* system. Or they can generate a movement by activating a motor cortex or nucleus such as the basal ganglia.

The appearance of an image in recall results from the reconstruction of a transient pattern (metaphorically, a map) in early sensory cortices, and the trigger for the reconstruction is the activation of dispositional representations elsewhere in the brain, as in the association cortex. The same type of mapped activation occurs in motor cortices and is the basis for movement. The dispositional representations on the basis of which movements occur are located in premotor cortices, basal ganglia, and limbic cortices. There is evidence that they activate both movements and internal images of body movement; because of the speedy nature of movements, the latter are often masked in consciousness by our awareness of the movement itself.

THOUGHT IS MADE LARGELY OF IMAGES

It is often said that thought is made of much more than just images, that it is made also of words and nonimage abstract symbols. Surely nobody will deny that thought includes words and arbitrary symbols. But what that statement misses is the fact that both words and arbitrary symbols are based on topographically organized representations and can become images. Most of the words we use in our inner speech, before speaking or writing a sentence, exist as auditory or visual images in our consciousness. If they did not become images, however fleetingly, they would not be anything we could know.[13] This is true even for those topographically organized representations that are not attended to in the clear light of consciousness, but are activated covertly. We know from priming experiments that although these representations are processed sub rosa, they can influence the course of the thought process, and even pop into consciousness a bit later. (Priming consists of activating a representation incompletely, or activating it but not attending to it).

We experience this phenomenon regularly. After a busy conversation involving several people, a word or statement that we did not hear during the conversation suddenly surfaces in our mind. We may be surprised by the fact that we missed it, how could we, and we may even check its reality, asking for instance, "Did you just say such and so?" Person X did indeed say such-and-so, but because you were concentrating on person Y, the mapped representations that were formed pertaining to what person X said were not attended to, and only a dispositional memory was made of it. As your concentration on person Y relaxed, and if the missed word or statement was relevant to you, the dispositional representation regenerated a topographically organized representation in an early sensory cortex; and since you were aware of it, it became an image. Note, by the way, that you never would have formed a dispositional representation without first forming a topographically mapped perceptual representation: there seems to be no anatomical way of getting complex sensory information into the association cortex that supports dispositional representations without first stopping in early sensory cortices. (This may not be true for noncomplex sensory information.)

The comments above apply as well to the symbols we may use in the mental solution of a mathematical problem (though perhaps not to all forms of mathematical thinking). If those symbols were not image-able, we would not know them and would not be able to manipulate them consciously. In this regard, it is interesting to observe that some insightful mathematicians and physicists describe their thinking as dominated by images. Often the images are visual, and they even can be somatosensory. Not surprisingly, Benoit Mandelbrot, whose life work is fractal geometry, says he always thinks in images.[14] He relates that the physicist Richard Feynman was not fond of looking at an equation without looking at the illustration that went with it (and note that both equation and illustration were images, in fact). As for Albert Einstein, he had no doubts about the process:

> The words or the language, as they are written or spoken, do not seem to play any role in my mechanism of thought. The psychi-cal entities which seem to serve as elements in thought are certain signs and more or less clear images which can be "volun-tarily" reproduced and combined. There is, of course, a certain connection between those elements and relevant logical con-cepts. It is also clear that the desire to arrive finally at logically connected concepts is the emotional basis of this rather vague play with the above mentioned elements.

Later in the same text he makes it even clearer:

> The above mentioned elements are, in my case, of visual and . . . muscular type. Conventional words or other signs have to Be sought for laboriously only in a secondary stage, when the mentioned associative play is sufficiently established and can be reproduced at will.[15]

The point, then, is that images are probably the main content of our thoughts, regardless of the sensory modality in which they are generated and regardless of whether they are about a thing or a process involving things; or about words or other symbols, in a given language, which correspond to a thing or process. Hidden behind those images, never or rarely knowable by us, there are indeed numerous processes that guide the generation and deployment of those images in space and time. Those processes utilize rules and strategies embodied in dispositional representations. They are *essential* for our thinking but are not a *content* of our thoughts.

The images that we reconstitute in recall occur side by side with the images formed upon stimulation from the exterior. The images reconstituted from the brain's interior are less vivid than those prompted by the exterior. They are "faint," as David Hume put it, in comparison with the "lively" images generated by stimuli from outside the brain. But they are images nonetheless.

SOME WORDS ON NEURAL DEVELOPMENT

As previously discussed, the brain's systems and circuits, as well as the operations they perform, depend on the pattern of connections among neurons and on the strength of the synapses constituting those connections. But how are the connection patterns and the synaptic strengths in our brains set, and when? Are they set at the same time for all systems throughout the brain? Once set, are they set forever? There are no definitive answers to these questions yet. Although knowledge on this subject is in constant flux, and not much should be taken for granted, things may work out like this:

1. The human genome (the sum total of the genes in our chromosomes) does not specify the entire structure of the brain. There are not enough genes available to determine the precise structure and place of everything in our organisms, least of all in the brain, where billions of neurons form their synaptic contacts. The disproportion is not subtle: we carry probably about 10^5 (100,000) genes, but we have more than 10^{15} (10 trillion) synapses in our brains. Moreover, the genet-

ically induced formation of tissues is assisted by interactions among cells, in which cell adhesion molecules and substrate adhesion molecules play an important role. What happens among cells, as development unfolds, actually controls, in part, the expression of the genes that regulate development in the first place. As far as one can tell, then, many structural specifics are determined by genes, but another large number can be determined only by the activity of the living organism itself, as it develops and continuously changes throughout its life span.[16]

2. The genome helps set the precise or nearly precise structure of a number of important systems and circuits in the evolutionarily old sectors of the human brain. Although we sorely need modern developmental studies concerned with these brain sectors, and although much could change as such studies materialize, the preceding statement seems reasonably certain for brain stem, hypothalamus, and basal forebrain, and quite likely for the amygdala and cingulate region. (I will say more about these structures and functions in the next chapters.) We share the essence of these brain sectors with individuals in numerous other species. The principal role of the structures in these sectors is to regulate basic life processes without recourse to mind and reason. The innate* patterns of activity of the neurons in these circuits do not generate images (although the consequences of their activity can be imaged); they regulate homeostatic mechanisms without which there is no survival. Without the innately set circuits of these brain sectors, we would not be able to

---

*Note that when I use the word innate (literally, present at birth), I am not excluding a role for environment and learning in the determination of a structure or pattern of activity. Nor am I excluding the potential for adjustments brought on by experience. I am using innate in the sense that William James used "pre-set," to refer to structures or patterns that are largely but not exclusively determined by the genome, and that are available to newborns to achieve homeostatic regulation.

breathe, regulate our heartbeat, balance our metabolism, seek food and shelter, avoid predators, and reproduce. Without this nuts-and-bolts biological regulation, individual and evolutionary survival would stop. Yet there is another role for these innate circuits which I must emphasize because it usually is ignored in the conceptualization of the neural structures supporting mind and behavior: *Innate circuits intervene not just in bodily regulation but also in the development and adult activity of the evolutionarily modern structures of the brain.*

3. The equivalent of the specifics that genes help set in the circuitry of the brain stem or hypothalamus comes to the remainder of the brain long after birth, as an individual develops through infancy, childhood, and adolescence, and as that individual interacts with the physical environment and other individuals. In all likelihood, as far as evolutionarily modern brain sectors are concerned, the genome helps set a general rather than a precise arrangement of systems and circuits. And how does the precise arrangement come about? It comes under *the influence of environmental circumstances complemented and constrained by the influence of the innately and precisely set circuits concerned with biological regulation.*

In short, the activity of circuits in the modern and experience-driven sectors of the brain (the neocortex, for example) is indispensable to produce a particular class of neural representations on which mind (images) and mindful actions are based. But the neocortex cannot produce images if the old-fashioned subterranean of the brain (hypothalamus, brain stem) is not intact and cooperative.

This arrangement may give one pause. Here we have innate circuits whose function is to regulate body function and to ensure the organism's survival, achieved by controlling the internal biochemical operations of the endocrine system, immune system, and viscera, and drives and instincts. Why should these circuits interfere with the shaping of the more modern and plastic ones concerned with representing our acquired experiences? The answer to this important question lies in the fact that both the records of experiences and the responses to them, if they are to be adaptive, must be evaluated and shaped by a fundamental set of preferences of the organism that consider survival paramount. It appears that because this evaluation and shaping are vital for the continuation of the organism, genes also specify that the innate circuits must exert a powerful influence on virtually the entire set of circuits that can be modified by experience. That influence is carried out in good part by "modulator" neurons acting on the remainder of the circuitry. These modulator neurons are located in the brain stem and the basal forebrain, and they are influenced by the interactions of the organism at any given moment. Modulator neurons distribute neurotransmitters (such as dopamine, norepinephrine, serotonin and acetylcholine) to widespread regions of the cerebral cortex and subcortical nuclei. This clever arrangement can be described as follows: (1) the innate regulatory circuits are involved in the business of organism survival and because of that they are privy to what is happening in the more modern sectors of the brain; (2) the goodness and badness of situations is regularly signaled to them; and (3) they express their inherent reaction to goodness and badness by influencing how the rest of the brain is shaped, so that it can assist survival in the most efficacious way.

Thus, as we develop from infancy to adulthood, the design of brain circuitries that represent our evolving body and its interaction with the world seems to depend on the activities in which the organism engages, and on the action of innate bioregulatory circuitries, *as the latter react to such activities*. This account underscores the inadequacy of conceiving brain, behavior, and mind in terms of nature versus nurture, or genes versus experience. Neither our brains nor our minds are tabulae rasae when we are born. Yet neither are they fully determined genetically. The genetic shadow looms large but is not complete. Genes provide for one brain component with *precise* structure, and for another component in which the precise structure is *to be determined*. But the to-be-determined structure can be achieved only under the influence of three elements: (1) the precise structure; (2) individual activity and circumstances (in which the final say comes from the human and physical environment as well as from chance); and (3) self-organizing pressures arising from the sheer complexity of the system. The unpredictable profile of experiences of each individual does have a say in circuit design, both directly and indirectly, via the reaction it sets off in the innate circuitries, and the consequences that such reactions have in the overall process of circuit shaping.[17]

I stated in chapter 2 that the operation of neuron circuits depends on the pattern of connections among the neurons and on the strength of the synapses that make those connections. In an excitatory neuron, for example, strong synapses facilitate firing, and weak synapses do the opposite. Now I can say that since different experiences cause synaptic strengths to vary within and across many neural systems, experience shapes the design of circuits. Moreover, in some systems more than in others, synaptic strengths can change throughout the life span, to reflect different organism experiences, and as a result, the design of brain circuits continues to change. The circuits are not only receptive to the results of first experience, but repeatedly pliable and modifiable by continued experiences.[18]

Some circuits are remodeled over and over throughout the life span, according to the changes an organism undergoes. Other circuits remain mostly stable and form the backbone of the notions we have constructed about the world within, and about the world outside. The idea that all circuits are evanescent makes little sense. Wholesale modifiability would have created individuals incapable of recognizing one another and lacking a sense of their own biography. That would not be adaptive, and clearly it does not happen. A simple proof that some acquired representations are relatively stable is found in the condition known as phantom limb. Some individuals who suffer the amputation of a limb (for instance the loss of the hand

and arm, leaving them with a stump above the level of the elbow)
report to their physicians that they still feel the missing limb in place,
that they can sense its imaginary movements, and that they can feel
pain or cold or warmth "in" the missing limb. Obviously these
patients possess a memory of their departed limb, or they would not
be able to form an image of it in their minds. Yet over time some
patients may experience a foreshortening of the phantom; apparently indicating that the memory—or its playback in consciousness—is undergoing revision.

The brain needs a balance between circuits whose firing alle-
giances may change like quicksilver, and circuits that are resistant
though not necessarily impervious to change. The circuits that help
us recognize our face in the mirror today, without surprise, have been
changed subtly to accommodate the structural modifications that
the time now spent has given those faces.

Six

Biological Regulation and Survival

DISPOSITIONS FOR SURVIVAL

AN ORGANISM'S SURVIVAL depends on a collection of biological processes that maintain the integrity of cells and tissues throughout its structure. Let me illustrate, albeit in a simplified way. Among many requirements, biological processes must have a proper supply of oxygen and nutrients, and that supply is based on respiration and feeding. For that purpose, the brain has innate neural circuits whose activity patterns, assisted by biochemical processes in the body proper, reliably control reflexes, drives, and instincts, and thus ensure that respiration and feeding are implemented as needed. To reflect back to the discussion in the previous chapter, the innate neural circuits contain dispositional representations. The activation of these dispositions sets in motion a complicated collection of responses.

On another front, to avoid destruction by predators or adverse environmental conditions, there are neural circuits for drives and instincts that cause, for example, fight or flight behaviors. Still other circuits control drives and instincts that help ensure the continuation of the individual's genes (through sexual behavior or care of kin). Numerous other specific circuits and drives might be mentioned, among them those related to the organism's seeking an ideal amount of light or darkness, heat or coolness, according to time of day or ambient temperature.

In general, drives and instincts operate either by generating a particular behavior directly or by inducing physiological states that lead individuals to behave in a particular way, mindlessly or not. Virtually all the behaviors ensuing from drives and instincts contribute to survival either directly, by performing a life-saving action, or indirectly, by propitiating conditions advantageous to survival or reducing the influence of potentially harmful conditions. Emotions and feelings, which are central to the view of rationality I am proposing, are a powerful manifestation of drives and instincts, part and parcel of their workings.

It would not be advantageous to allow the dispositions controlling basic biological processes to change much. A significant change would bring with it the risk of major malfunction in varied organ systems and the prospect of a disease state or even death. This is not to deny that we can willfully influence the behaviors that usually are driven by those innate neural patterns. We can hold our breath as we swim underwater, for a stretch; we can decide to go on a prolonged fast; we can influence our heart rate, easily, and even alter our systemic blood pressure, not so easily. But in none of these instances is there evidence that dispositions change. What changes is one component or another of the ensuing behavioral pattern, which we succeed in inhibiting in a number of ways, be it through muscular force (holding our breath by contracting the upper airway and rib cage) or sheer willpower. Nor is it to deny that the innate patterns can be modulated in their firing—made more likely to fire or not—by neural signals from other brain regions, or by chemical signals, such as hormones and neuropeptides, brought to them in the bloodstream or through axons. In fact, many neurons throughout the brain have receptors for hormones, such as those from the reproductive, adrenal, and thyroid glands. Both early development and regular operation of those circuitries are influenced by such signaling.

Some of the basic regulatory mechanisms operate at covert level and are never directly knowable to the individual inside whom they operate. You do not know the state of the various circulating hormones, potassium ions, or the number of red blood cells in your body unless you assay it. But slightly more complex regulatory mechanisms, involving overt behaviors, let you know about their existence, indirectly, when they drive you to perform (or not) in a particular way. These are called instincts.

Instinctual regulation might be explained in a simplified way by this example: Several hours after a meal your blood sugar level drops, and neurons in the hypothalamus detect the change; activation of the pertinent innate pattern makes the brain alter the body state so that the probability for correction can be increased; you feel hungry, and initiate actions to end your hunger; you eat, and the ingestion of food brings about a correction in blood sugar; finally, the hypothalamus again detects a change in blood sugar, this time an increase, and the appropriate neurons place the body in the state whose experience constitutes the feeling of satiety.

The goal of the entire process was saving your body. The signal to initiate the process came from your body. The signals that entered your consciousness, in order to force you to save your body, also came from your body. As the cycle concluded, the signals that told you that your body was no longer in danger came from your body. You might say that this is government for the body and by the body, although it is sensed and managed by the brain.

Such regulatory mechanisms ensure survival by driving a disposition to excite some pattern of body changes (a drive), which can be a body state with a specific meaning (hunger, nausea), or a recognizable emotion (fear, anger), or some combination thereof. The excitement can be triggered from the "visceral" inside (low blood sugar in the internal milieu), from the outside (a threatening stimulus), or from the "mental" inside (realization that a catastrophe is about to happen). Each of these can engage an internal bioregulatory response, or an instinctual behavior pattern, or a newly created action plan, or any or all of them. The basic neural circuitries that operate this entire cycle are standard equipment for your organism, as much as the brakes are in a car. You did not have to have them specially installed. They constitute a "preorganized mechanism"—a notion to which I will return in the next chapter. All you had to do was tune the mechanism to your environment.

Preorganized mechanisms are important not just for basic biological regulation. They also help the organism classify things or events as "good" or "bad" because of their possible impact on survival. In other words, the organism has a basic set of preferences—or criteria, biases, or values. Under their influence and the agency of experience, the repertoire of things categorized as good or bad grows rapidly, and the ability to detect new good and bad things grows exponentially.

If a given entity out in the world is a component of a scene in which one *other* component was a "good" or "bad" thing, that is, excited an innate disposition, the brain may classify the entity for which no value had been innately preset as if it too is valuable, whether or not it is. The brain extends special treatment to that entity simply because it is close to one that is important for sure. You may call this reflected glory, if the new entity is close to a good thing, or guilt by association, if it is close to a bad one. The light that shines on a bona fide important item, good or bad, will shine also on its company. What the brain must do to operate in this fashion is come into the world with considerable "innate knowledge" about how to regulate itself and the rest of the body. As the brain incorporates dispositional representations of interactions with entities and scenes relevant for innate regulation, it increases the chances of including entities and scenes that may or may not be directly relevant to survival. And as this happens, our growing sense of whatever the world outside may be, is apprehended as a modification in the neural space in which body and brain interact. It is not only the separation between mind and brain that is mythical: the separation between mind and body is probably just as fictional. The mind is embodied, in the full sense of the term, not just embrained.

MORE ON BASIC REGULATION

The innate neural patterns that seem most critical for survival are maintained in circuits of the brain stem and hypothalamus. The latter is a key player in the regulation of the endocrine glands— among them the pituitary, the thyroid, the adrenals, and the reproductive organs, all of which produce hormones—and in the function of the immune system. Endocrine regulation, which depends on chemical substances released into the bloodstream rather than on neural impulses, is indispensable to maintaining metabolic function and managing the defense of biological tissues against micropredators such as viruses, bacteria, and parasites.[1]

Biological regulation related to the brain stem and hypothalamus is complemented by controls in the limbic system. This is not the place to discuss the intricate anatomy and detailed function of this sizable brain sector, but it should be noted that the limbic system participates also in the enactment of drives and instincts and has an especially important role in emotions and feelings. I suspect that unlike the brain stem and hypothalamus, however, whose circuitry is mostly innate and stable, the limbic system contains both innate circuitry and circuitry modifiable by the experience of the ever-evolving organism.

With the help of nearby structures in the limbic system and brain stem, the hypothalamus regulates the *internal milieu* (the term and concept, which I have used before, are inherited from the pioneer biologist Claude Bernard), which you may picture as all the biochemical processes occurring in an organism at any given moment. Life depends on those biochemical processes' being kept within a suitable range, since excessive departures from that range, at key points in the composite profile, may result in disease or death. In turn, the hypothalamus and interrelated structures are regulated not only by neural and chemical signals from other brain regions, but also by chemical signals arising in various body systems.

This chemical regulation is especially complex, as the following will indicate: The production of hormones released by the thyroid and adrenal glands, without which we cannot live, is controlled partly by chemical signals from the pituitary gland. The pituitary is itself controlled partly by chemical signals released from the hypothalamus into the bloodstream near the pituitary, and the hypothalamus is controlled partly by neural signals from the limbic system and, indirectly, from the neocortex. (Consider the significance of the following observation: The abnormal electrical activity of certain limbic system circuits during seizures causes not only an abnormal mental state but also profound hormonal abnormalities which can lead to a host of body diseases such as ovarian cysts.) In return, each hormone in the bloodstream acts on the gland that secreted it, as well as on the pituitary, the hypothalamus, and other brain sectors. In other words, neural signals give rise to chemical signals, which give rise to other chemical signals, which can alter the function of many cells and tissues (including those in the brain), and alter the regulatory circuits that initiated the cycle itself. These many nested regulatory mechanisms manage body conditions locally and globally so that the organism's constituents, from molecules to organs, operate within the parameters required for survival.

The layers of regulation are interdependent along many dimensions. A given mechanism may, for instance, depend on a simpler mechanism, and be influenced by a more complex or equally complex mechanism. Activity in the hypothalamus can influence neocortical activity, directly or via the limbic system, and the reverse is also true.

Consequently, as might be expected, there is a documented brain-body interaction, and we may glean perhaps less visible mind-body interactions. Consider the following example: Chronic mental stress, a state related to processing in numerous brain systems at the

level of neocortex, limbic system, and hypothalamus, seems to lead
to overproduction of a chemical, calcitonin gene-related peptide, or
CGRP, in nerve terminals within the skin.[2] As a result, CGRP
excessively coats the surface of Langerhans cells, an immune-
related cell whose job it is to capture infectious agents and deliver
them to lymphocytes so that the immune system can counteract
their presence. If completely coated by CGRP, the Langerhans cells
are disabled and can no longer perform their guardian function. The
end result is that the body is more vulnerable to infection, now that a
major entryway is less well defended. And there are other examples
of mind-body interaction: Sadness and anxiety can notably alter the
regulation of sexual hormones, causing not only changes in sexual
drive but also variations in menstrual cycle. Bereavement, again a
state dependent on brainwide processing, leads to a depression of
the immune system such that individuals are more prone to infection
and, whether as a direct result or not, more likely to develop certain
types of cancer.[3] One *can* die of a broken heart.

The reverse influence, that of chemical substances from the body
on the brain, has been observed as well, of course. It is no surprise
that tobacco, alcohol, and drugs (medical and nonmedical) enter the
brain and modify its function, and thus alter the mind. Some of
the actions of body chemicals fall directly over neurons or their
support systems; some are indirect, via the neurotransmitter media-
tor neurons located in the brain stem and basal forebrain, which
were discussed previously. Upon firing, those small collections of
neurons can deliver a dose of dopamine, norepinephrine, serotonin,
or acetylcholine to widespread regions of the of the brain including
the cerebral cortex and basal ganglia. The arrangement might be
imagined as a set of well-engineered sprinkler devices, each delivering its chemical substance to particular systems and, within the
systems, to particular circuits with particular types and amounts of
receptors.[4] Changes in the amount and distribution of release of one
of those transmitters, or even changes in the relative balance of
transmitters at a particular site, can influence cortical activity
rapidly and profoundly and give rise to states of depression or ela-

tion, even mania. (See chapter 7.) Thought processes can slow down or speed up; the profusion of recalled images can decrease or increase; the creation of novel combinations of images can be enhanced or shut down. The ability to concentrate on a particular mind content fluctuates accordingly.

TRISTAN, ISOLDE, AND THE LOVE POTION

Remember the story of Tristan and Isolde? The plot revolves around a transformation in the relation between the two protagonists. Isolde asks her maid, Brangäne, to prepare a death potion, but instead Brangäne prepares a "love potion," which both Tristan and Isolde drink, not knowing what it is supposed to produce. The mysterious drink unleashes the deepest possible passion in them, and draws them to each other in a rapture that nothing can break—not even the fact that each of them on their own is wretchedly betraying the benevolent King Mark. Richard Wagner captured the force of the lovers' bond in perhaps the most exalted and desperate love passages in the history of music, in his opera *Tristan und Isolde*. One has to wonder why he was attracted to this story, and why millions have, for more than a century, communed with his rendition of it.

The answer to the first question is that the composition celebrated a very real and similar passion in Wagner's life. Wagner and Mathilde Wesendonk had fallen in love, entirely against their soundest judgment, when one considers that she was the wife of his generous benefactor and that he was already married. Wagner did have a sense for the concealed and undetainable forces that may overpower one's will and which, for lack of more suitable explanations, have been attributed to magic or to destiny.

The answer to the second question is more tantalizing. There are indeed potions in our own bodies and brains, capable of forcing on us behaviors that we may or may not be unable to suppress by strong resolution. A key example is the chemical substance oxytocin.[5] In the case of mammals, humans included, it is manufactured both in the brain (in the supraoptic and parvoventral nuclei of the hypothalamus) and in the body (in the ovary or in the testes). It can be released by the brain in order to participate, for instance, directly or by interposed hormones, in the regulation of metabolism; or it can be released by the body, during childbirth, sexual stimulation of genitals or nipples, or orgasm, when it acts not only on the body itself (by relaxing muscles during childbirth, for instance), but also in the brain. What it can do there is nothing short of the effect of legendary elixirs. In general, it influences a whole range of grooming, locomotion, sexual, and maternal behaviors. More important, for my story, it facilitates social interactions and induces bonding between mating partners. A good example comes from Thomas Insel's studies on the prairie vole, a rodent with gorgeous fur. After their lightning courtship and a first day of repeated and intense copulation, the male and female remain inseparable till death does them part. The male actually acquires a sour disposition toward any creature other than his beloved and is usually quite helpful around the nest. Such bonding is not only a charming adaptation but a most advantageous one, in many species, since it keeps together those who must rear the offspring, and it also helps with other aspects of social organization. Humans certainly use many of oxytocin's effects all the time, although they have learned to avoid, under certain circumstances, those effects which may or may not be ultimately good. Remember that the love potion was not good for Wagner's Tristan and Isolde. Three hours later, not counting the intermissions, they die a desolate death.

To the neurobiology of sex, about which a lot is currently known, we can now add the beginnings of the neurobiology of attachment, and, armed with both, throw a bit more light on that complex set of mental states and behaviors we call love.

What is at play here, in the massively recurrent circuit arrangements I have outlined, is a collection of feedforward and feedback loops in which some of the loops are purely chemical. Perhaps most significant about this arrangement is the fact that the brain structures involved in basic biological regulation are also part of the regulation of behavior and are indispensable to the acquisition and normal function of cognitive processes. The hypothalamus, the brain stem, and the limbic system intervene in body regulation *and* in all neural processes on which mind phenomena are based, for example, perception, learning, recall, emotion and feeling, and—as I shall propose later—reasoning and creativity. Body regulation, survival, and mind are intimately interwoven. The interweaving occurs in biological tissue and uses chemical and electrical signaling, all within Descartes' *res extensa* (the physical realm in which he includes the body and the surrounding environment but not the nonphysical soul, which belongs to the *res cogitans*). Curiously, it happens most strongly not far from the pineal gland, inside which Descartes once sought to imprison the nonphysical soul.

BEYOND DRIVES AND INSTINCTS

How much drives and instincts alone can ensure an organism's survival seems to depend on the complexity of the environment and the complexity of the organism in question. Among animals, from insects to mammals, there are unequivocal examples of successful coping with particular forms of environment on the basis of innate strategies, and no doubt those strategies often include complex aspects of social cognition and behavior. I never cease to marvel at the intricate social organization of our distant monkey cousins, or at the elaborate social observances of so many birds. When we consider our own species, however, and the far more varied and largely unpredictable environments in which we have thrived, it is apparent that we must rely on highly evolved genetically based biological mechanisms, as well as on suprainstinctual survival strategies that have developed in society, are transmitted by culture, and require, for their application, consciousness, reasoned deliberation, and willpower. This is why human hunger, desire, and explosive anger do not proceed unchecked toward feeding frenzy, sexual assault, and murder, at least not always, assuming that a healthy human organism has developed in a society in which the suprainstinctual survival strategies are actively transmitted and respected.

Western and Eastern thinkers, religious and not, have been aware of this for millennia; closer to us, the topic preoccupied both Descartes and Freud, to name but two. The control of animal inclination by thought, reason, and the will was what made us human, according to Descartes' *Passions of the Soul*.[6] I agree with his formulation, except that where he specified a control achieved by a nonphysical agent I envision a biological operation structured within the human organism and not one bit less complex, admirable, or sublime. The creation of a superego which would accommodate instincts to social dictates was Freud's formulation, in *Civilization and Its Discontents*, which was stripped of Cartesian dualism but was nowhere explicit in neural terms.[7] A task that faces neuroscientists today is to consider the neurobiology supporting adaptive supraregulations, by which I mean the study and understanding of the brain structures required to know about those regulations. I am not attempting to reduce social phenomena to biological phenomena, but rather to discuss the powerful connection between them. It should be clear that although culture and civilization arise from the behavior of biological individuals, the behavior was generated in collectives of individuals interacting in specific environments. Culture and civilization could not have arisen from single individuals and thus cannot be reduced to biological mechanisms and, even less, can they be reduced to a subset of genetic specifications. Their comprehension demands not just general biology and neurobiology but the methodologies of the social sciences as well.

In human societies there are social conventions and ethical rules over and above those that biology already provides. Those additional layers of control shape instinctual behavior so that it can be adapted flexibly to a complex and rapidly changing environment and ensure survival for the individual and for others (especially if they belong to the same species) in circumstances in which a preset response from the natural repertoire would be immediately or eventually counterproductive. The perils preempted by such conventions and rules may be immediate and direct (physical or mental harm), or remote and indirect (future loss, embarrassment). Although such conventions and rules need be transmitted only through education and socialization, from generation to generation, I suspect that the neural representations of the wisdom they embody, and of the means to implement that wisdom, are inextricably linked to the neural representation of innate regulatory biological processes. I see a "trail" connecting the brain that represents one, to the brain that represents the other. Naturally, that trail is made up of connections among neurons.

For most ethical rules and social conventions, regardless of how elevated their goal, I believe one can envision a meaningful link to simpler goals and to drives and instincts. Why should this be so? Because the consequences of achieving or not achieving a rarefied social goal contribute (or are perceived as contributing), albeit indirectly, to survival and to the quality of that survival.

Does this mean that love, generosity, kindness, compassion, honesty, and other commendable human characteristics are nothing but the result of conscious but selfish, survival-oriented neurobiological regulation? Does this deny the possibility of altruism and negate free will? Does this mean that there is no true love, no sincere friendship, no genuine compassion? That is definitely *not* the case. Love is true, friendship sincere, and compassion genuine, if I do not lie about how I feel, if I *really* feel loving, friendly, and compassionate. Perhaps I would be more eligible for praise if I arrived at such sentiments by means of pure intellectual effort and willpower, but what if I have not, what if my current nature helps me get there faster, and be nice and honest without even trying? The truth of the feeling (which concerns how what I do and say matches what I have in mind), the magnitude of the feeling, and the beauty of the feeling, are not endangered by realizing that survival, brain, and proper education have a lot to do with the reasons why we experience such feelings. The same applies to a considerable extent to altruism and free will. Realizing that there are biological mechanisms behind the most sublime human behavior does not imply a simplistic reduction to the nuts and bolts of neurobiology. In any case, the partial explanation of complexity by something less complex does not signify debasement.

The picture I am drawing for humans is that of an organism that comes to life designed with automatic survival mechanisms, and to which education and acculturation add a set of socially permissible and desirable decision-making strategies that, in turn, enhance survival, remarkably improve the quality of that survival, and serve as the basis for constructing a *person*. At birth, the human brain comes to development endowed with drives and instincts that include not just a physiological kit to regulate metabolism but, in addition, basic devices to cope with social cognition and behavior. It emerges from child development with additional layers of survival strategy. The neurophysiological base of those added strategies is interwoven with that of the instinctual repertoire, and not only modifies its use but extends its reach. The neural mechanisms supporting the suprainstinctual repertoire may be similar in their overall formal design to those governing biological drives, and may be constrained by them. Yet they require the intervention of society to become whatever they become, and thus are related as much to a given culture as to general neurobiology. Moreover, out of that dual constraint, suprainstinctual survival strategies generate something probably unique to humans: a moral point of view that, on occasion, can transcend the interests of the immediate group and even the species.

Seven

Emotions
and Feelings

H ow does one translate into neurobiological terms the ideas presented at the end of the previous chapter? The evidence on biological regulation demonstrates that response selections of which organisms are not conscious and which are thus not deliberated take place continuously in evolutionarily old brain structures. Organisms whose brains only include those archaic structures and are devoid of evolutionarily modern ones—reptiles, for instance—operate such response selections without difficulty. One might conceptualize the response selections as an elementary form of decision making, provided it is clear that it is not an aware self but a set of neural circuits that is doing the deciding.

Yet it is also well accepted that when social organisms are confronted by complex situations and are asked to decide in the face of uncertainty, they must engage systems in the neocortex, the evolutionarily modern sector of the brain. There is evidence for a relation between the expansion and subspecialization of the neocortex, and the complexity and unpredictability of environments with which such expansion permits individuals to cope. Relevant in this regard is John Allman's valuable finding that, independently of body size, the neocortex of fruit-eating monkeys is larger than that of leaf-eating monkeys.[1] Fruit-eating monkeys must have a richer memory so that they can remember when and where to look for edible fruit lest they encounter fruitless trees or rotten fruit. Their larger neocortices support the greater factual memory capacity they require.

So blatant is the discrepancy between the processing capacities of "low and old" and "high and new" brain structures that it has fostered an implicit and seemingly sensible view on the respective responsibilities of those brain sectors. In simple terms: The old brain core handles basic biological regulation down in the basement, while up above the neocortex deliberates with wisdom and subtlety. Upstairs in the cortex there is reason and willpower, while downstairs in the subcortex there is emotion and all that weak, fleshy stuff.

This view, however, does not capture the neural arrangement that underlies rational decision-making as I see it. For one thing, it is not compatible with the observations discussed in part I. For another, there is evidence that longevity, a likely reflection of the quality of reasoning, is correlated not only with increased size of the neocortex, as expected, but also with increased size of the hypothalamus, the main compartment of the downstairs.[2] The apparatus of rationality, traditionally presumed to be *neo*cortical, does not seem to work without that of biological regulation, traditionally presumed to be *sub*cortical. Nature appears to have built the apparatus of rationality not just on top of the apparatus of biological regulation, but also *from* it and *with* it. The mechanisms for behavior beyond drives and instincts use, I believe, both the upstairs and the downstairs: the neocortex becomes engaged *along with* the older brain core, and rationality results from their concerted activity.

A question may arise here about the degree to which rational and nonrational processes are aligned respectively with cortical and subcortical structures in the human brain. To approach this question, I now turn to emotion and feeling, central aspects of biological regulation, to suggest that they provide the bridge between rational and nonrational processes, between cortical and subcortical structures.

EMOTIONS

About a century ago, William James, whose insights on the human mind have been rivaled only by Shakespeare's and Freud's, produced a truly startling hypothesis on the nature of emotion and feeling. Consider his words:

> If we fancy some strong emotion and then try to abstract from our consciousness of it all the feelings of its bodily symptoms, we find we have nothing left behind, no "mind-stuff" out of which the emotion can be constituted, and that a cold and neutral state of intellectual perception is all that remains.

Using compelling illustrations, James went on to state:

> What kind of an emotion of fear would be left if the feeling neither of quickened heart-beats nor of shallow breathing, neither of trembling lips nor of weakened limbs, neither of goose-flesh nor of visceral stirrings, were present, it is quite impossible for me to think. Can one fancy the state of rage and picture no ebullition in the chest, no flushing of the face, no dilatation of the nostrils, no clenching of the teeth, no impulse to vigorous action, but in their stead limp muscles, calm breathing, and a placid face?[3]

With these words, well ahead of both his time and ours, I believe William James seized upon the mechanism essential to the understanding of emotion and feeling. Unfortunately, and uncharacteristically for him, the rest of his proposal fell so short of the variety and complexity of the phenomena it addressed, that it has been the source of endless and sometimes hopeless controversy.[4] (I cannot do justice here to the extensive scholarship on this subject, which has been reviewed by George Mandler, Paul Ekman, Richard Lazarus, and Robert Zajonc.)

The main problem some have had with James's view is not so much his stripping emotion down to a process that involved the body, of all possible things, shocking as that must have been to his critics, but rather that he gave little or no weight to the process of evaluating mentally the situation that causes the emotion. His account works well for the first emotions one experiences in life, but it does not do justice to what Othello goes through in his mind before he develops jealousy and anger, or to what Hamlet broods about before exciting his body into what he will perceive as disgust, or to the twisted reasons why Lady Macbeth should experience ecstasy as she leads her husband into a murderous rampage.

Almost as problematic was the fact that James made no provision for an alternative or supplementary mechanism to generate the feeling that corresponds to a body excited by emotion. In the Jamesian view, the body is *always* interposed in the process. Moreover, James had little to say about the possible roles of emotion in cognition and behavior. As I suggested in the Introduction, however, emotions are not a luxury. They play a role in communicating meanings to others, and they may also play the cognitive guidance role that I propose in the next chapter.

In short, James postulated a basic mechanism in which particular stimuli in the environment excite, by means of an innately set and inflexible mechanism, a specific pattern of body reaction. There was no need to evaluate the significance of the stimuli in order for the reaction to occur. Matters were not made more clear by his lapidary statement: "Every object that excites an instinct excites an emotion as well."

In many circumstances of our life as social beings, however, we know that our emotions are triggered only after an evaluative, voluntary, nonautomatic mental process. Because of the nature of our experience, a broad range of stimuli and situations has become associated with those stimuli which are innately set to cause emotions. The reaction to that broad range of stimuli and situations can be filtered by an interposed mindful evaluation. And because of the thoughtful, evaluative filtering process, there is room for variation in the extent and intensity of preset emotional patterns; there is, in effect, a modulation of the basic machinery of the emotions gleaned by James. Moreover, there seem to be other neural means to achieve the body sense that James considered the essence of the emotional process.

In the pages ahead I outline my views on emotion and feeling. I begin with the perspective of personal history, and clarify the differences between the emotions we experience early in life, for which a Jamesian "preorganized mechanism" would suffice, and the emotions we experience as adults, whose scaffolding has been built gradually on the foundation of those "early" emotions. I propose calling "early" emotions primary, and "adult" emotions secondary.

Primary Emotions

To what degree are emotional reactions wired in at birth? I would say that neither animals nor humans are, of necessity, innately wired for bear fear, or eagle fear (although some animals and humans may be wired for spider fear and snake fear). One possibility I have no problem with is that we are wired to respond with an emotion, in preorganized fashion, when certain features of stimuli in the world or in our bodies are perceived, alone or in combination. Examples of such features include size (as in large animals); large span (as in flying eagles); type of motion (as in reptiles); certain sounds (such as growling); certain configurations of body state (as in the pain felt during a heart attack). Such features, individually or conjunctively, would be processed and then detected by a component of the brain's limbic system, say, the amygdala; its neuron nuclei possess a dispositional representation which triggers the enactment of a body state characteristic of the emotion fear, and alters cognitive processing in a manner that fits the state of fear (we will see further on that the brain can "simulate" body states and bypass the body, and we will discuss how the cognitive alteration is achieved). Note that in order to cause a body response, one does not even need to "recognize" the bear, or snake, or eagle, as such, or to know what, precisely, is causing pain. All that is required is that early sensory cortices detect and categorize the key feature or features of a given entity (e.g., animal, object), and that structures such as the amygdala receive

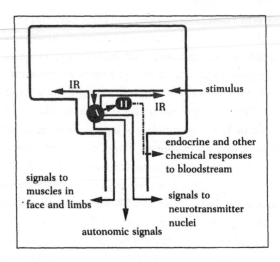

Figure 7-1. Primary Emotions. The black perimeter stands for the brain and brain stem. After an appropriate stimulus activates the amygdala (A), a number of responses ensue: internal responses (marked IR); muscular responses; visceral responses (autonomic signals); and responses to neurotransmitter nuclei and hypothalamus (H). The hypothalamus gives rise to endocrine and other chemical responses which use a bloodstream route. I am leaving out of the diagram several other brain structures required to implement this large array of responses. For instance, the muscular responses with which we express emotions, say, in body posture, probably utilize structures in the basal ganglia (namely, the so-called ventral striatum).

signals concerning their *conjunctive* presence. A baby chick in a nest does not know what eagles are, but promptly responds with alarm and by hiding its head when wide-winged objects fly overhead at a certain speed. (See Fig. 7-1.)

By itself, the emotional response can accomplish some useful goals: speedy concealment from a predator, for instance, or display of anger toward a competitor. The process does not stop with the bodily changes that define an emotion, however. The cycle continues, certainly in humans, and its next step is the *feeling of the emotion* in connection to the object that excited it, the realization of the nexus between object and emotional body state. Now, it may be asked, why would anyone need to become cognizant of such a relation? Why complicate matters and bring consciousness into this process, if there is already a means to respond adaptively at an automated level? The answer is that consciousness buys an enlarged protection policy. Consider this: If you come to *know* that animal or object or situation X causes fear, you will have two ways of behaving toward X. The first way is innate; you do not control it. Moreover, it is not specific to X; a large number of creatures, objects, and circumstances can cause the response. The second way is based on your own experience and is specific to X. Knowing about X allows you to think ahead and predict the probability of its being present in a given environment, so that you can avoid X, preemptively, rather than just have to react to its presence in an emergency.

But there are other advantages of "feeling" your emotional reactions. You can generalize your knowledge, and decide, for example, to be cautious with anything that looks like X. (Of course, if you overgeneralize and behave overcautiously, you may become phobic—which is not so good.) Furthermore, you may have discovered, in the course of your first encounter with X, something peculiar and potentially vulnerable in X's behavior. You may want to exploit that vulnerability in your next encounter, and that is one more reason why you need to have *known*. In short, feeling your emotional states, which is to say being conscious of emotions, offers you *flexibility of response based on the particular history of your interactions with the environment*. Although you need innate devices to start the ball of knowledge rolling, feelings offer you something extra.

Primary emotions (read: innate, preorganized, Jamesian) depend on limbic system circuitry, the amygdala and anterior cingulate being the prime players. Evidence that the amygdala is the key player in preorganized emotion comes from observations in both animals and humans. The amygdala has been the precise focus of various animal studies by Pribram, Weiskrantz, Aggleton and Passingham, and more recently, and perhaps most comprehensively, by Joseph LeDoux.[5] Other contributions to the field include those of E.T. Rolls, Michael

Davis, and of Larry Squire and his group, whose work, although aimed at understanding memory, also revealed a connection between the amygdala and emotion.[6] The amygdala was also implicated in emotion by Wilder Penfield and by Pierre Gloor and Eric Halgren when they studied epileptic patients whose surgical evaluation required electrical stimulation of varied regions in the temporal lobe.[7] More recently there have been supporting observations on the human amygdala by investigators in my group and in retrospect, the first hint that amygdala and emotion might be related can be found in the work of Heinrich Kluver and Paul Bucy,[8] who showed that surgical resection of the part of the temporal lobe containing the amygdala created affective indifference, among a variety of other symptoms. (For evidence on the relation between anterior cingulate and emotion, see Chapter 4 of this book, and pertinent descriptions by Laplane et al., 1981, and A. Damasio and Van Hoesen, 1983.[9])

But the mechanism of primary emotions does not describe the full range of emotional behaviors. They are, to be sure, the basic mechanism. However, I believe that in terms of an individual's development they are followed by mechanisms of *secondary emotions*, which occur once we begin experiencing feelings and forming *systematic connections between categories of objects and situations, on the one hand, and primary emotions, on the other*. Structures in the limbic system are not sufficient to support the process of secondary emotions. The network must be broadened, and it requires the agency of prefrontal and of somatosensory cortices.

Secondary Emotions

To address the notion of secondary emotions let us shift to an example drawn from an adult's experience. Imagine meeting a friend whom you have not seen in a long time, or being told of the unexpected death of a person who worked closely with you. In either real instance—and perhaps even as you imagine the scenes now—you experience an emotion. What happens to you, neurobiologically, as that emotion occurs? What does it really mean to "experience an emotion"?

If I were there when you imagined either of those scenes, or similar ones, I might be able to make some observations. After forming mental images of key aspects in the scenes (the encounter with the long lost friend; the death of a colleague), there is a change in your body state defined by several modifications in different body regions. If you meet an old friend (in your imagination), your heart may race, your skin may flush, the muscles in your face change around the mouth and eyes to design a happy expression, and muscles elsewhere will relax. If you hear of an acquaintance's death, your heart may pound, your mouth dry up, your skin blanch, a section of your gut contract, the muscles in your neck and back tense up while those in your face design a mask of sadness. In either case, there are changes in a number of parameters in the function of viscera (heart, lungs, gut, skin), skeletal muscles (those that are attached to your bones), and endocrine glands (such as the pituitary and adrenals). A number of peptide modulators are released from the brain into the bloodstream. The immune system also is modified rapidly. The baseline activity of smooth muscles in artery walls may increase, and produce contraction and thinning of blood vessels (the result is pallor); or decrease, in which case the smooth muscle would relax and blood vessels dilate (the result is flushing). As a whole, the set of alterations defines a profile of departures from a range of average states corresponding to functional balance, or homeostasis, within which the organism's economy operates probably at its best, with lesser energy expenditure and simpler and faster adjustments. This range of functional balance should not be seen as static; it is a continuous succession of profile changes within upper and lower limits, in constant motion. It might be likened to the condition of a waterbed when someone walks on it in varied directions: some areas are depressed, while others rise; ripples form; the entire bed is modified as a whole, but the changes are within a range specified by the physical limits of the unit: a boundary containing a certain amount of fluid.

In your hypothetical experience of emotion, many parts of your body are placed in a new state, one in which significant changes are introduced. What happens in the organism to effect such changes?

1. The process begins with the conscious, deliberate considerations you entertain about a person or situation. These considerations are expressed as mental images organized in a thought process, and they concern myriad aspects of your relationship with the given person, reflections on the current situation and its consequences for you and others, in sum, a cognitive evaluation of the contents of the event of which you are a part. Some of the images you conjure up are nonverbal (the likeness of a given person in a given place), while others are verbal (words and sentences regarding attributes, activities, names, and so on). The neural substrate for such images is a collection of separate topographically organized representations, occurring in varied early sensory cortices (visual, auditory, and others). Those representations are constructed under the guidance of dispositional representations held in distributed manner over a large number of higher-order association cortices.

2. At a nonconscious level, networks in the prefrontal cortex automatically and involuntarily respond to signals arising from the processing of the above images. This prefrontal response comes from dispositional representations that embody knowledge pertaining to how certain types of situations usually have been paired with certain emotional responses, in your individual experience. In other words, it comes from *acquired* rather than *innate* dispositional representations, although, as discussed previously, the acquired dispositions are obtained under the influence of dispositions that are innate. What the acquired dispositional representations embody is your unique experience of such relations in your life. Your experience may be at subtle or at major variance with that of others; it is yours alone. Although the relations between type of situation and emotion are, to a great extent,

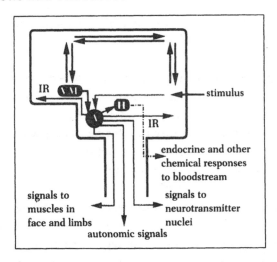

Figure 7-2. Secondary Emotions. The stimulus may still be processed directly via the amygdala but is now also analyzed in the thought process, and may activate frontal cortices (VM). VM acts via the amygdala (A). In other words, secondary emotions utilize the machinery of Primary Emotions. Again, I am deliberately oversimplifying, since numerous prefrontal cortices other than VM are also activated, but I believe the essence of the mechanism is as shown in the diagram. Note how VM depends on A to express its activity, how it is piggy-backed on it, so to speak. This dependence-precedence relationship is a good example of nature's tinkering style of engineering. Nature makes use of old structures and mechanisms in order to create new mechanisms and obtain new results.

similar among individuals, unique, personal experience customizes the process for every individual. To summarize: The prefrontal, acquired dispositional representations needed for secondary emotions are a separate lot from the innate dispositional representations needed for primary emotions. But as you will discover below, the former need the latter in order to express themselves.

3. Nonconsciously, automatically and involuntarily, the response of the prefrontal dispositional representations described in the preceding paragraph is signaled to the amygdala and the anterior cingulate. Dispositional represen-

tations in the latter regions respond (a) by activating nuclei of the autonomic nervous system and signaling to the body via peripheral nerves, with the result that viscera are placed in the state most commonly associated with the type of triggering situation; (b) by dispatching signals to the motor system, so that the skeletal muscles complete the external picture of an emotion in facial expressions and body posture; (c) by activating the endocrine and peptide systems, whose chemical actions result in changes in body and brain states; and finally, (d) by activating, with particular patterns, the nonspecific neurotransmitter nuclei in brain stem and basal forebrain which then release their chemical messages in varied regions of the telencephalon (e.g., basal ganglia and cerebral cortex). This apparently exhausting collection of actions is a massive response; it is varied. It is aimed at the whole organism, and in a healthy person, it is a marvel of coordination.

The changes caused by (a), (b), and (c) impinge on the body, cause an "emotional body state," and are subsequently signaled back to the limbic *and* somatosensory systems. The changes caused by (d), which do not arise in the body proper but rather in a group of brain stem structures in charge of body regulation, have a major impact in the style and efficiency of cognitive processes, and constitute a parallel route for the emotional response. The different effects of (a), (b), and (c), on the one hand, and (d), on the other, will become clearer in the discussion of feelings (see below).

It now should be clear that the emotional processing impaired in patients with prefrontal damage is of the secondary type. These patients cannot generate emotions relative to the images conjured up by certain categories of situation and stimuli, and thus cannot have the ensuing feeling. This is borne out in clinical observations and special tests, described in chapter 9. Those same prefrontal patients can have primary emotions, however, and that is why their affect may appear to be intact at first glance (they would show fear if someone screamed unexpectedly right behind them, or if their houses shook in an earthquake). On the contrary, patients with limbic system damage in the amygdala or anterior cingulate usually have a more pervasive impairment of both primary and secondary emotions, and thus are more recognizably blunted in their affect.

Nature, with its tinkerish knack for economy, did not select independent mechanisms for expressing primary and secondary emotions. It simply allowed secondary emotions to be expressed by the same channel already prepared to convey primary emotions.

I see the *essence* of emotion as the collection of changes in body state that are induced in myriad organs by nerve cell terminals, under the control of a dedicated brain system, which is responding to the content of thoughts relative to a particular entity or event. Many of the changes in body state—those in skin color, body posture, and facial expression, for instance—are actually perceptible to an external observer. (Indeed, the etymology of the word nicely suggests an external direction, from the body: *emotion* signifies literally "movement out.") Other changes in body state are perceptible only to the owner of the body in which they take place. But there is more to emotion than its essence.

In conclusion, emotion is the combination of a *mental evaluative process*, simple or complex, with *dispositional responses to that process*, mostly *toward the body proper*, resulting in an emotional body state, but also *toward the brain itself* (neurotransmitter nuclei in brain stem), resulting in additional mental changes. Note that, for the moment, I leave out of emotion the perception of all the changes that constitute the emotional response. As you will soon discover, I reserve the term *feeling* for the experience of those changes.

The Specificity of Neural Machinery Behind the Emotions

The specificity of the neural systems dedicated to emotion has been established from studies of focal brain damage. As I see it, damage to the limbic system impairs the processing of primary emotion; damage to prefrontal cortices compromises the processing of secondary emotion. An intriguing neural correlate of human emotion was established by Roger Sperry and his collaborators, among them Joseph Bogen, Michael Gazzaniga, Jerre Levy, and Eran Zaidel: structures in the human right cerebral hemisphere have a preferential involvement in the basic processing of emotion.[10] Other investigators, namely, Howard Gardner, Kenneth Heilman, Joan Borod, Richard Davidson, and Guido Gainotti, have added supporting evidence in favor of right-hemisphere dominance for emotion.[11] Current research in my laboratory generally supports the idea of asymmetry in the process of emotion, but also indicates that the asymmetries do not pertain to all emotions equally.

The degree of neural specificity of the systems dedicated to emotion can be gauged by considering the impairment of its expression. When a stroke destroys the motor cortex on the brain's left hemisphere and, as a result, the patient has paralysis on the right side of the face, the muscles cannot act and the mouth tends to be pulled toward the normally moving side. Asking the patient to open the mouth and reveal the teeth only heightens the asymmetry. Yet when the patient smiles or laughs spontaneously, in response to a humorous remark, something entirely different happens: the smile is normal, both sides of the face move as they should, and the expression is natural, no different from the usual pre-paralysis smile of that individual. This illustrates that the motor control for an emotion-related movement sequence is *not* in the same location as the control for a voluntary act. The emotion-related movement is triggered elsewhere in the brain, even if the arena for the movement, the face and its musculature, is the same. (See fig. 7-3.)

If you study a patient in whom a stroke has damaged the anterior cingulate in the left hemisphere, you will see precisely the opposite result. In repose or in emotion-related movement, the face is asymmetrical, less mobile on the right than on the left. But if the patient tries to contract the facial muscles willfully, the movements are carried out normally and symmetry returns. Emotion-related movement, then, is controlled from the anterior cingulate region, from other limbic cortices (in the medial temporal lobe), and from the

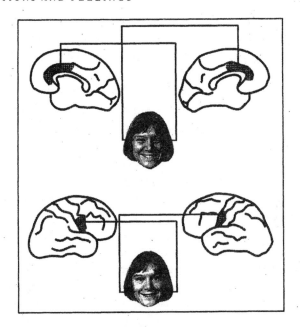

Figure 7-3. The neural machinery for the control of face musculature in the "true" smile of an emotional situation (top panels) is different from the machinery for voluntary (nonemotional) control of the same musculature (bottom panels). The true smile is controlled from limbic cortices and probably uses the basal ganglia for its expression.

basal ganglia, regions whose damage or dysfunction yields a so-called reverse or emotional facial paralysis.

My mentor Norman Geschwind, the Harvard neurologist whose work bridged the classical and modern eras of brain and mind research in humans, was fond of pointing out that the reason we have difficulty smiling naturally for photographers (the "say cheese" situation) is that they ask us to control our facial muscles willfully, using the motor cortex and its pyramidal tract. (The pyramidal tract is the massive set of axons that arises in the primary motor cortex, area 4 of Brodmann, and descends to innervate the nuclei in the brain stem and spinal cord that control voluntary motion through peripheral nerves.) We thus produce, as Geschwind liked to call it, a "pyramidal smile." We cannot mimic easily what the anterior cingulate can achieve effortlessly; we have no easy neural route to exert volitional control over the anterior cingulate. In order to smile "naturally," you have only a few options: learn to act, or get somebody to tickle you or tell you a good joke. The career of actors and politicians hinges on this simple, annoying disposition of neurophysiology.

The problem has long been recognized by professional actors, and has led to different acting techniques. Some, well exemplified by Laurence Olivier's, rely on skillfully creating, under volitional control, a set of movements that credibly suggest emotion. Drawing on detailed knowledge of what emotions (their expressions) look like to the external observer, and on the memory of how one usually feels as such external changes occur, the great actors of that tradition fake it, with great determination. That few succeed is a measure of the hurdles brain physiology poses for them.

Another technique, exemplified by the Lee Strasberg–Elia Kazan "Method" acting (inspired by the work of Konstantin Stanislavsky), relies on having actors generate an emotion, create the real thing rather than simulate it. This can be more convincing and engaging, but it requires special talent and maturity to rein in the automated processes unleashed by the real emotion.

The difference between facial expressions of genuine and make-believe emotions was first noted by Charles Darwin in *The Expression of the Emotions in Man and Animals*, published in 1872.[12] Darwin was aware of observations made a decade earlier by Guillaume-Benjamin Duchenne about the musculature involved in smiling and the type of control needed to move that musculature.[13] Duchenne determined that a smile of real joy required the combined involuntary contraction of two muscles, the zygomatic major and the orbicularis oculi. (See fig. 7-4.) He discovered further that the latter muscle could be moved only involuntarily; there was no way of activating it willfully. The involuntary activators of the orbicularis oculi, as Duchenne put it, were "the sweet emotions of the soul." As for the zygomatic major, it can be activated both involuntarily and by our will and is thus the proper avenue for smiles of politeness.

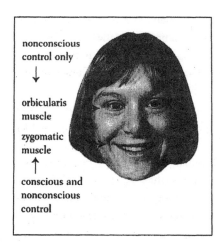

nonconscious
control only

↓

orbicularis
muscle

zygomatic
muscle

↑

conscious and
nonconscious
control

Figure 7-4. Nonconscious and
conscious control of the face
musculature.

FEELINGS

What is a feeling? Why do I not use the terms "emotion"
and "feeling" interchangeably? One reason is that although
some feelings relate to emotions, there are many that do not:
all emotions generate feelings if you are awake and alert, but not
all feelings originate in emotions. I call background feelings those
that do not originate in emotions and I discuss them later in the
chapter.

I shall begin by considering the *feelings of emotions,* and for that I
will return to your emotional state in the example discussed above.
All the changes that an external observer can identify and many
others that an observer cannot, such as a heart beating faster or a
contracted gut, you perceived *internally.* All these changes are being
signaled continuously to the brain through nerve terminals that
bring to it impulses from skin, blood vessels, viscera, voluntary
muscles, joints, and so on. In neural terms, the return leg of this trip
depends on circuits that originate in the head, neck, trunk, and
limbs, course in the spinal cord and brain stem toward the reticular
formation (a collection of brain stem nuclei involved in the control of
wakefulness and sleep, among other functions) and thalamus, and

travel on to the hypothalamus, limbic structures, and several distinct somatosensory cortices in the insular and parietal regions. The latter cortices in particular receive an account of what is happening in your body, moment by moment, which means that they get a "view" of the ever-changing landscape of your body during an emotion. If you recall the waterbed image, you can conceive of that view as continuous signaling representing many of the local changes in the bed, the up-and-down movements it undergoes as someone walks on it. In the cerebral cortices that receive those signals continuously, there is an ever-changing pattern of neural activity. There is nothing static about it, no baseline, no little man—the homunculus—sitting in the brain's penthouse like a statue, receiving signals from the corresponding part of the body. Instead there is change, ceaseless change. Some of the patterns are organized topographically, some less so, and they are not found in one single map, at one single center. There are many maps, coordinated by mutually interactive neuron connections. (Whatever the metaphor we use to illustrate the point, it is important to realize that *current* body representations do not occur within a rigid cortical map as decades of human brain diagrams have insidiously suggested. They occur as a dynamic, newly instantiated, "on-line" representation of what is happening in the body now. Their value resides with that freshness and "on-lineness," so well demonstrated in the work of Michael Merzenich previously cited.)

In addition to the "neural trip" of your emotional state back to the brain, your organism also used a parallel "chemical trip." Hormones and peptides released in the body during the emotion can reach the brain via the bloodstream, and penetrate the brain actively, through the so-called blood–brain barrier or, even more easily, through brain regions lacking that barrier (e.g., the area postrema) or having devices that signal to varied parts of the brain (e.g., the subfornical organ). Not only can the brain construct, in some of its systems, a multifarious neural view of the body landscape that other brain systems have induced, but the construction of the view itself, as well as its use, can be influenced by the body directly (think of oxytocin, discussed in chapter 6). What gives the body landscape its character at a given moment is not just a set of neural signals but also a set of chemical signals that modify the mode in which neural signals are processed. Think of this as *the* reason why certain chemical substances have played a major role in so many cultures; and consider that the drug problem that our society currently faces—and I refer to both illegal and legal drugs—cannot be solved without understanding in depth the neural mechanisms we are discussing here.

As body changes take place, you get to know about their existence and you can monitor their continuous evolution. You perceive changes in your body state and follow their unfolding over seconds and minutes. That process of continuous monitoring, that experience of what your body is doing *while* thoughts about specific contents roll by, is the essence of what I call a feeling. (fig. 7-5) If an emotion is a collection of changes in body state connected to particular mental images that have activated a specific brain system, *the essence of feeling an emotion is the experience of such changes in juxtaposition to the mental images that initiated the cycle.* In other words, a feeling depends on the juxtaposition of an image of the body proper to an image of something else, such as the visual image of a face or the auditory image of a melody. The substrate of a feeling is

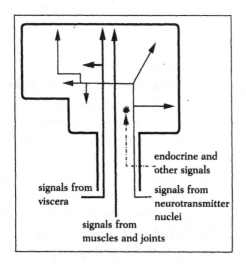

signals from viscera

signals from neurotransmitter nuclei

endocrine and other signals

signals from muscles and joints

Figure 7-5. To feel an emotion it is necessary *but not sufficient that neural signals from viscera, from muscles and joints, and from neurotransmitter nuclei—all of which are activated during the process of emotion—reach certain subcortical nuclei and the cerebral cortex. Endocrine and other chemical signals also reach the central nervous system via the bloodstream among other routes.*

completed by the changes in cognitive processes that are simultaneously induced by neurochemical substances (for instance, by neurotransmitters at a variety of neural sites, resulting from the activation
in neurotransmitter nuclei which was part of the initial emotional
response).*

At this point I must make two qualifications. The first concerns
the notion of "juxtaposition" in the definition above. I chose this
term because I think the image of the body proper appears *after* the
image of the "something else" has been formed and held active, and
because the two images remain separate, neurally, as I suggested in
the section on images in chapter 5. In other words, there is a
"combination" rather than a "blending." It might be appropriate to
use the term *superposition* for what seems to happen to the images of
body proper and "something else" in our integrated experience.

The idea that the "qualified" (a face) and the "qualifier" (the
juxtaposed body state) are combined but not blended helps explain
why it is possible to feel depressed even as one thinks about people or
situations that in no way signify sadness or loss, or feel cheerful for
no immediately explainable reason. The qualifier states may be
unexpected and sometimes unwelcome. Their psychological motivation may be unapparent or nonexistent, the process arising in a
psychologically neutral physiological change. Neurobiologically
speaking, the unexplainable qualifiers affirm the relative autonomy
of the neural machinery behind the emotions. But they also remind
us of the existence of a vast domain of nonconscious processes,
some part of which is amenable to psychological explanation and
some part of which is not.

The essence of sadness or happiness is the combined perception
of certain body states with whatever thoughts they are juxtaposed to,
complemented by a modification in the style and efficiency of the

---

*The definitions of "emotion" and "feeling" presented here are not orthodox. Other
authors often use these words interchangeably, or "feeling" may not be used at all
and "emotion" divided into expressive and experienced components. Implying separate terms might help further investigation of these phenomena.

thought process. In general, because both the signal of the body state (positive or negative) and the style and efficiency of cognition were triggered from the same system, they tend to be concordant. (Although the concordances between body-state signal and cognitive style can be broken down in normal as well as in pathologic states.) Along with negative body states, the generation of images is slow, their diversity small, and reasoning inefficient; along with positive body states the generation of images is rapid, their diversity wide, and reasoning may be fast though not necessarily efficient. When negative body states recur frequently, or when there is a sustained negative body state, as happens in a depression, the proportion of thoughts which are likely to be associated with negative situations does increase, and the style and efficiency of reasoning suffer. The sustained elation of manic states produces the opposite result. In *Darkness Visible*, the memoir of his own depression, William Styron has provided definitive descriptions of such a condition. He writes about its essence as a tormenting sense of pain ". . . most closely connected to drowning or suffocation—but even these images are off the mark." But he does not miss the description of the accompanying state of his cognitive processes: "Rational thought was usually absent from my mind at such times, hence *trance*. I can think of no more apposite word for this state of being, a condition of helpless stupor in which cognition was replaced by that 'positive and active anguish.'" (Positive and active anguish were the terms used by William James to describe his own depression.)

The other qualification: I have provided my view of what the essential constituents of a feeling may be, cognitively and neurally; only further inquiry will tell whether this view is correct. But I have not explained *how* we feel a feeling. Receiving a comprehensive set of signals about the body state in the appropriate brain regions is the necessary beginning but is not sufficient for feelings to be felt. As I suggested in the discussion on images, a further condition for the experience is a correlation of the ongoing representation of the body with the neural representations constituting the self. A feeling about a particular object is based on the subjectivity of the perception of the object, the perception of the body state it engenders, and the perception of modified style and efficiency of the thought process as all of the above happens.

Fooling the Brain

What evidence is there supporting the claim that body states cause feelings? Some evidence comes from neuropsychological studies correlating loss of feeling with damage to the brain regions necessary to represent body states (see chapter 5), but studies conducted in normal individuals are also telling in this respect as well, specifically those by Paul Ekman.[14] When he gave normal experimental subjects instructions on how to move their facial muscles, in effect "composing" a specific emotional expression on the subjects' faces without their knowing his purpose, the result was that the subjects experienced a feeling appropriate to the expression. For instance, a roughly and incompletely composed happy facial expression led to the subjects' experiencing "happiness," an angry facial expression to their experiencing "anger," and so on. This is impressive if we consider that the subjects could perceive only sketchy, fragmentary facial postures, and that since they were neither perceiving nor evaluating any real situation that might trigger an emotion, their bodies were not exhibiting, at the outset, the visceral profile that accompanies a certain emotion.

Ekman's experiment suggests either that a fragment of the body pattern characteristic of an emotional state is enough to produce a feeling of the same signal, or that the fragment subsequently triggers the rest of the body state and that leads to the feeling. Curiously, not all parts of the brain are fooled, as it were, by a set of movements that is not produced through the usual means. New evidence from electrophysiological recordings shows that make-believe smiles generate different patterns of brain waves from those generated by real smiles.[15] At first glance the electrophysiological finding may seem to contradict that of the previously cited experiment, but it does not: although they reported the feeling appropriate to the fragment of facial expression, the subjects were well aware that they were not happy or angry at any particular thing. We cannot fool ourselves any more than we can fool others when we only smile politely, and that is what the electrical recording seems to correlate with so nicely. This may also be the very good reason why great actors, opera singers, and others manage to survive the simulation of exalted emotions they regularly put themselves through, without losing control.

I asked Regina Resnik, the most memorable operatic Carmen and Clytemnestra of our time, and the veteran of a thousand nights of musical anger and madness, how difficult it had been to remain separate from the exorbitant emotions of her characters. Not difficult at all, said she, once she learned the secrets of her technique. Nobody would have guessed, watching and hearing her, that she was just bodily "portraying" emotion rather than "feeling" it. But she does admit that once, playing in Tchaikovsky's *The Queen of Spades*, alone on the dark stage for the death-by-fright scene of the Old Countess, she did become one with her character and was terrified.

VARIETIES OF FEELINGS

As indicated at the beginning of the chapter, there are many varieties of feelings. The first variety is based on emotions, the most universal of which are Happiness, Sadness, Anger, Fear, and Disgust, and correspond to profiles of body state response which are largely preorganized in the James sense. When the body conforms to the profiles of one of those emotions we *feel* happy, sad, angry, fearful, disgusted. When we have feelings connected with emotions, attention is allocated substantially to body signals, and parts of the body landscape move from the background to the foreground of our attention.

A second variety of feelings is based on emotions that are subtle variations of the five mentioned above: euphoria and ecstasy are variations of happiness; melancholy and wistfulness are variations of sadness; panic and shyness are variations of fear. This second variety of feelings is tuned by experience, when subtler shades of cognitive

Varieties of Feelings
Feelings of Basic Universal Emotions
Feelings of Subtle Universal Emotions
Background Feelings

state are connected to subtler variations of emotional body state. It is the connection between an intricate cognitive content and a variation on a preorganized body-state profile that allows us to experience shades of remorse, embarrassment, *Schadenfreude*, vindication, and so on.[16]

Background Feelings

But I am postulating another variety of feeling which I suspect preceded the others in evolution. I call it *background feeling* because it originates in "background" body states rather than in emotional states. It is not the Verdi of grand emotion, nor the Stravinsky of intellectualized emotion but rather a minimalist in tone and beat, the feeling of life itself, the sense of being. I hope the notion may be helpful in the future analysis of the physiology of feelings.

More restricted in range than the emotional feelings described previously, background feelings are neither too positive nor too negative, although they can be perceived as mostly pleasant or unpleasant. In all probability it is these feelings, rather than emotional ones, that we experience most frequently in a lifetime. We are only subtly aware of a background feeling, but aware enough to be able to report instantly on its quality. A background feeling is not what we feel when we jump out of our skin for sheer joy, or when we are despondent over lost love; both of these actions correspond to emotional body states. A background feeling corresponds instead to the body state prevailing *between* emotions. When we feel happiness, anger, or another emotion, the background feeling has been superseded by an emotional feeling. The background feeling is our image of the body landscape when it is not shaken by emotion. The concept of "mood," though related to that of background feeling, does not exactly capture it. When background feelings are persistently of the same type over hours and days, and do not change quietly as thought contents ebb and flow, the collection of background feelings probably contributes to a mood, good, bad, or indifferent.

If you try for a moment to imagine what it would be like to be *without* background feelings, you will have no doubt about the notion I am introducing. I submit that without them the very core of your representation of self would be broken. Let me explain why I think so.

As I have indicated, the representations of current body states occur in multiple somatosensory cortices in the insula and parietal regions, and also in the limbic system, hypothalamus, and brain stem. These regions, in both left and right hemispheres, are coordinated by neuron connections, the right hemisphere dominating over the left. Much remains to be discovered about the precise connectional specifications of this system (regrettably it is one of the least-studied sectors of the primate brain), but this much seems clear: A composite, ongoing representation of current body states is distributed over a large number of structures in both subcortical and cortical locations. A good part of the input from visceral states ends up in structures that might be called "nonmapped," although plenty of visceral input is mapped well enough for us to detect pain or discomfort in identifiable areas of the trunk or limbs. While it is true that the maps we make for the viscera are less precise than those we make for the outside world, the alleged vagueness and instances of mapping error have been exaggerated, largely by invoking such phenomena as "referred pain" (e.g., feeling pain in the left arm or abdomen during a myocardial infarction, or pain underneath the right scapula when the gallbladder is inflamed). As for the input from muscles and joints, it ends up in topographically mapped structures.

In addition to "on-line," dynamic body maps, there are somewhat more stable maps of general body structure, which probably represent proprioception (muscular and joint sense) and interoception (visceral sense), and which constitute the basis for our notion of body image. Those representations are "off-line," or dispositional, but they can be activated into the topographically organized somatosensory cortices, side by side with the on-line representation of the body states *now,* to provide an idea of what our bodies *tend to be like,* rather than what they are now. The best evidence for this kind of representation is the phenomenon of phantom limb, mentioned earlier. After a surgical amputation, some patients imagine the missing limb as if it were still there. They are even capable of perceiving imaginary modifications in the state of the nonexistent limb, such as a particular motion, pain, temperature, and so on. My interpretation of this phenomenon is that in the absence of on-line input from the missing limb, there prevails the on-line input from a dispositional representation of that limb: that is, the reconstruction through the process of recall of a previously acquired memory.

Those who believe that little of the body state appears in consciousness under normal conditions may want to reconsider. It is true that we are not aware of every part of our body, all of the time, because representations of external events, through vision, hearing, or touch, as well as internally generated images, effectively distract us from the ongoing, uninterruptible representation of the body. But the fact that our focus of attention is usually elsewhere, where it is most needed for adaptive behavior, does not mean the body representation is absent, as you can easily confirm when the sudden onset of pain or minor discomfort shifts the focus back to it. The background body sense is continuous, although one may hardly notice it, since it represents not a specific part of anything in the body but rather an overall state of most everything in it. Yet such an ongoing, unstoppable representation of the body state is what allows you to reply promptly to the specific question "How do you *feel*?" with an answer that does relate to whether you feel fine or do not feel that well. (Note that the question is not the simple "How are you?" to which one may reply politely and perfunctorily without saying anything about one's body state.) The background state of the body is monitored continuously, and thus it is intriguing to wonder what would happen if, all of a sudden, it were to disappear; if, when asked how you felt, you found you knew nothing about that background state; if, when your leg hurt and you deliberately uncrossed it, the momentary discomfort were an isolated percept set loose in your mind, rather than part of the sense of a body whose wholeness you have easy access to. It is known for certain that even the much simpler, relatively circumscribed suspension of proprioception, which can be caused by a disease of peripheral nerves, creates a profound disruption of mental processes. (Oliver Sacks has written an evocative description of one such patient.[17]) It is to be expected, then, that a more pervasive loss or modification of the overall sense of body state will produce an even greater disturbance, and that is indeed the case.

As described in chapter 4, some patients with prototypical and complete anosognosia become unaware of their general medical condition. They do not know that they are suffering from the invariably devastating results of some major illness, most often a stroke, or a brain tumor arising in the brain itself or secondary to cancer elsewhere in the body. They do not recognize that they are paralyzed, although they will concur that their left limbs do not move, when they are confronted with the fact and forced to see, for instance, their motionless left hand and arm. They cannot picture the consequences of their medical situation and are not concerned with their future. Their emotional display is restrained or nonexistent, and their feelings—by their own admission and from an observer's inference—are correspondingly flat.

The pattern of brain damage in such anosognosics results in the disruption of cross-talk among regions involved in body-state mapping, and often in the destruction of some of those regions themselves. The regions are all in the right hemisphere, although they receive input from both right and left sides of the body. The key regions are in the insula, the parietal lobe, and the white matter containing connections among them and, in addition, connections to and from thalamus, to and from frontal cortex, and to basal ganglia.

Using the notion of background feeling I can now indicate what I think happens in anosognosia. Unable to avail themselves of current body input, anosognosics fail to update the representation of their bodies and as a result fail to recognize, through the somatosensory system, promptly and automatically, that the reality of their body landscape has changed. They still can form in their minds an image of what their bodies were like, an image that is now outdated. And since their body was fine, that is what they venture to report.

Patients with the phantom-limb condition may report that they feel their missing limb is still there, but they realize that it clearly is not. They do not have a delusion or hallucination; indeed, it is their sense of reality that leads them to complain about their inconvenient state. But anosognosics have no automatic reality check. Either because the condition involves information about most of the body, rather than a part, or because it involves visceral information more than any other, or for both reasons, they are different. The lack of updated body signals leads not only to irrational reports about their motor defects, but also to inappropriate emotion and feeling relative to their state of health. These patients appear unconcerned about their condition, some being inappropriately jocular, others monotonously sullen. When forced to reason about their state, on the basis of new facts presented through other channels, verbally or through direct visual confrontation, they momentarily acknowledge their new situation, but the realization is soon forgotten. Somehow, what does not come naturally and automatically through the primacy of feeling cannot be maintained in the mind.

Patients with anosognosia offer us a view of a mind deprived of the possibility of sensing *current* body state, especially as it concerns background feeling. I suggest that these patients' self, unable to plot current body signals on the ground reference of the body, is no longer integral. Knowledge about personal identity is still available and retrievable in language form: anosognosics remember who they are, where they live and worked, who the people close to them are. But that wealth of information cannot be used to reason effectively on the current personal and social state. The theory that these patients construct of their own minds and of the minds of others is woefully, irrevocably out-of-date, out of step with the historical time that they and their observers are immersed in.

The continuity of background feelings befits the fact that the living organism and its structure are continuous as long as life is maintained. Unlike our environment, whose constitution does change, and unlike the images we construct relative to that environment, which are fragmentary and conditioned by external circumstance, background feeling is mostly about body states. Our individual identity is anchored on this island of illusory living sameness against which we can be aware of myriad other things that manifestly change around the organism.

THE BODY AS THEATER FOR THE EMOTIONS

One of the criticisms leveled at William James concerns the idea that we always use the body as theater for the emotions. Although I believe that in many situations emotions and feelings are operated precisely in that manner, from mind/brain to body, and back to mind/brain, I believe also that in numerous instances the brain learns to concoct the fainter image of an "emotional" body state, without having to reenact it in the body proper. Moreover, as we have previously discussed, the activation of neurotransmitter nuclei in brain stem and their responses bypass the body, although, in a most curious way, the neurotransmitter nuclei are part and parcel of the brain representation of body regulation. There are thus neural devices that help us feel "as if" we were having an emotional state, as if the body were being activated and modified. Such devices permit us to bypass the body and avoid a slow and energy-consuming process.

We conjure up some semblance of a feeling within the brain alone. I doubt, however, that those feelings feel the same as the feelings freshly minted in a real body state.

"As if" devices would have been developed while we were growing up and adapting to our environment. The association between a certain mental image and the surrogate of a body state would have been acquired by repeatedly associating the images of given entities or situations with the images of freshly enacted body states. To have a particular image trigger the "bypass device," it was first necessary to run the process through the body theater, to loop it through the body, as it were. (See Fig. 7-6.)

Why should "as if" feelings feel different? Let me illustrate at least one reason why I think so: Picture the situation of a normal person connected to a polygraph, a laboratory instrument that permits assessment of the shape and magnitude of emotional reactions in the form of continuous graphs. Now imagine the person participating in a psychological experiment during which the examiner will consider

Figure 7-6. A diagram of the "body loop" and of the "as if" loop. In both body loop and "as if" loop panels, the brain is represented by the top black perimeter and the body by the bottom one. The processing in the "as if" loop bypasses the body entirely.

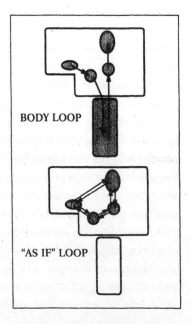

BODY LOOP

"AS IF" LOOP

certain responses correct and deserving of some kind of reward, or incorrect and deserving of some degree of penalty. The subject, upon being told that a particular move he or she made in the experiment is correct and is being rewarded, generates a response, which appears as a curve with a particular onset and rise shape and a particular top magnitude. Sometime later, another move by the subject brings on a penalty and that also generates a response, but this time the shape of the curve is quite different and it rises higher than in the previous one. A bit later another move triggers a stiffer penalty, and not only is the response curve different, but the recording needle careens across the paper and almost jumps off the recording surface.

The meaning of this difference in responses is well known: Different degrees of reward and punishment cause different reactions, mentally and bodily, and the polygraph records the bodily reaction. There is, however, disagreement about the relation between body reaction and mind reaction. From my perspective, regular feeling comes from a "readout" of the body changes. But we must consider an alternative view, that the body indeed is changed by the emotional reaction, but that the feeling does not necessarily come from that change; that the same brain agent that sets body changes in motion informs another brain site, presumably the somatosensory system, of the type of change being commissioned from the body. According to this alternative view, feelings would come directly from the latter set of signals, which thus would be processed entirely within the brain, although there would still be concomitant body changes. The point, for those who espouse this view, is that the body changes occur in parallel with the feelings rather than being causative of the feelings. Feelings would always derive from the "as-if loop" device, which would not be a supplement to the basic "body loop" device, as I proposed above, but rather the essential mechanism of feeling.

Why do I find the alternative view less satisfactory than mine? For one thing, an emotion is not induced by neural routes alone. There is also the chemical route. The sector of the brain that induces the emotion may signal the neural component of the induction within itself, to another sector of itself, but it is not likely to signify the chemical component in the same manner. Moreover, the brain is not likely to predict how all the commands—neural and chemical, but especially the latter—will play out in the body, because the play-out and the resulting states depend on local biochemical contexts and on numerous variables within the body itself which are not fully represented neurally. What is played out in the body is constructed anew, moment by moment, and is not an exact replica of anything that happened before. I suspect that the body states are not algorithmically predictable by the brain, but rather that the brain waits for the body to report what actually has transpired.

The alternative view of emotions and feelings would be limited, time after time, to a fixed repertoire of emotion/feeling patterns, which would not be modulated by the real-time, real-life conditions of the organism at any one moment. These patterns might be helpful if that were all we had to go on, but they would still be "rebroadcasts" rather than "live performances."

The brain probably cannot predict the exact landscapes the body will assume, after it unleashes a barrage of neural and chemical signals on the body, no more than it can predict all the imponderables of a specific situation as it unfolds in real life and real time. Whether for an emotional state or a nonemotional background state, the body landscape is always new and hardly ever stereotyped. If all of our feelings were of the "as if" type, we would have no notion of the ever-changing modulation of affect that is such a salient trait of our mind. Anosognosia suggests that the normal mind requires a steady flow of updated information from body states. It might be that, as currently designed, the brain needs an affirmation of our living state before it cares to keep itself awake and aware.

MINDING THE BODY

It does not seem sensible to leave emotions and feelings out of any overall concept of mind. Yet respectable scientific accounts of cognition do precisely that, by failing to include emotions and feelings in their treatment of cognitive systems. This is an omission to which I alluded in the Introduction: emotions and feelings are considered elusive entities, unfit to share the stage with the tangible contents of the thoughts they nonetheless qualify. This strict view, which excludes emotion from mainstream cognitive science, has a counterpart in the no less traditional brain-sciences view to which I alluded earlier in this chapter; namely, that emotions and feelings arise in the brain's down-under, in as subcortical as a subcortical process can be, while the stuff that those emotions and feelings qualify arises in the neocortex. I cannot endorse these views. First, it is apparent that emotion is played out under the control of both subcortical and neocortical structures. Second, and perhaps more important, *feelings are just as cognitive as any other perceptual image,* and just as dependent on cerebral-cortex processing as any other image.

To be sure, feelings are about something different. But what makes them different is that they are first and foremost about the body, that they offer us *the cognition of our visceral and musculoskeletal state* as it becomes affected by preorganized mechanisms and by the cognitive structures we have developed under their influence. Feelings let us *mind the body,* attentively, as during an emotional state, or faintly, as during a background state. They let us mind the body "live," when they give us perceptual images of the body, or "by rebroadcast," when they give us recalled images of the body state appropriate to certain circumstances, in "as if" feelings.

Feelings offer us a glimpse of what goes on in our flesh, as a momentary image of that flesh is juxtaposed to the images of other objects and situations; in so doing, feelings modify our comprehensive notion of those other objects and situations. By dint of juxtaposition, body images give to other images a *quality* of goodness or badness, of pleasure or pain.

I see feelings as having a truly privileged status. They are represented at many neural levels, including the neocortical, where they are the neuroanatomical and neurophysiological equals of whatever is appreciated by other sensory channels. But because of their inextricable ties to the body, they come first in development and retain a primacy that subtly pervades our mental life. Because the brain is the body's captive audience, feelings are winners among equals. And since what comes first constitutes a frame of reference for what comes after, feelings have a say on how the rest of the brain and cognition go about their business. Their influence is immense.

THE PROCESS OF FEELING

What are the neural processes by which we *feel* an emotional state or a background state? I do not know precisely; I think I have the beginning of the answer, but I am not certain about the ending. The question of how we feel rests on our understanding of consciousness, something about which it pays to be modest, and that is not the subject of this book. We can still ask the question, however, and disqualify those answers which cannot possibly work, and consider where some answers might be found in the future.

One answer that is falsely satisfactory has to do with the neurochemistry of emotion. Discovering the chemicals involved in emotions and moods is not enough to explain how we feel. It has long been known that chemical substances can change emotions and moods; alcohol, narcotics, and a host of pharmacological agents can modify how we feel. The well-known relationship between chemistry and feeling has prepared scientists and the public for the discovery that the organism produces chemicals that can have similar effect. The idea that endorphins are the brain's own morphine and can easily change how we feel about ourselves, about pain, and about the world is now well accepted. So is the idea that the neurotransmitters dopamine, norepinephrine, and serotonin, as well as peptide neuromodulators, can have similar effects.

It is important to realize, however, that knowing that a given chemical (manufactured inside or outside the body) causes a given feeling to occur is not the same as knowing the mechanism for how this result is achieved. Knowing that a substance is working on certain systems, in certain circuits and receptors, and in certain neurons, does not explain *why* you feel happy or sad. It establishes a working relationship among the substance, the systems, the circuits,

the receptors, the neurons, and the feeling, but it does not tell you
how you get from one to the other. It is only the beginning of an
explanation. If feeling happy or sad corresponds in good part to a
change in the neural representation of ongoing body states, then the
explanation requires that the chemicals act on the sources of those
neural representations, that is, the body proper itself, and the many
levels of neural circuitry whose activity patterns represent the body.
Of necessity, understanding the neurobiology of feeling requires the
understanding of the latter. If feeling happy or sad also corresponds
in part to the cognitive modes under which your thoughts are operating, then the explanation also requires that the chemical acts on the
circuits which generate and manipulate images. Which means that
reducing depression to a statement about the availability of serotonin or norepinephrine in general—a popular statement in the days
and age of Prozac—is unacceptably rude.

Another falsely satisfactory answer is the simple equation of feeling with the neural representation of what is happening in the body
landscape at a given moment. Regrettably this is not enough; we
must discover how the constantly and properly modulated body
representations become subjective, how they become part of the self
that owns them. How can we explain such a process neurobiologically, without resorting to the convenient tale of the homunculus
perceiving the representation?

Beyond the neural representation of the body state, then, I see a
need to posit at least two major components in the neural mechanisms underlying feeling. The first, which would occur early in the
process, is described below. The second, which is anything but
straightforward, has to do with the self, and is taken up in chapter 10.

In order for us to feel a certain way about a person or an event, the
brain must have a means to represent the causal link between
the person or event and the body state, preferably in an unequivocal
manner. In other words, you do not want to connect an emotion,
positive or negative, to the wrong person or thing. We often make
wrong connections, for instance, when we associate a person, object, or place with a bad turn of events, but some of us try to keep

from making those erroneous links. Superstition is based on this sort of spurious causal association: a hat on a bed brings bad luck, as does a black cat crossing your path; walk under a ladder, you'll meet with misfortune; and so on. When the spurious alignment of emotion (fear) and object is pervasive, phobic behavior will ensue. (The flip side of phobic behavior is just as annoying. By overassociating positive emotions with people, objects, or places, too often and indiscriminately, we may feel more positive and relaxed about many situations than we should, and may end up like Pollyanna.)

This sense of precise cause-and-effect may arise from activity in convergence zones that perform a mutual brokerage between body signals and signals about the entity causing the emotion. Convergence zones operate as "third-party" brokers by means of the reciprocal feedforward and feedback connections they maintain with their sources of input. The players in my proposed arrangement are an explicit representation of the *causative entity*; an explicit representation of the *current body state*; and a *third-party representation*. In other words, the brain activity that signals a certain entity and transiently forms a topographically organized representation in the appropriate early sensory cortices; the brain activity that signals body-state changes and transiently forms a topographically organized representation in early somatosensory cortices; and a representation, located in a convergence zone, that receives signals from those first two sites of brain activity, by feedforward neural connections. This third-party representation preserves the order of the onset of brain activity, and in addition maintains activity and attentional focus by means of feedback connections to the two sites of brain activity. Signals among the three players lock the ensemble in relatively synchronous activity, for a brief period. In all likelihood, this process requires cortical and subcortical structures, namely those in the thalamus.

Emotion and feeling thus rely on two basic processes: (1) the view of a certain body state juxtaposed to the collection of triggering and

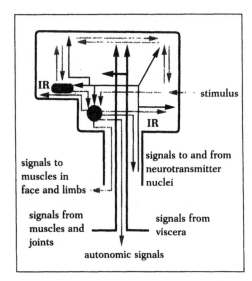

stimulus

signals to and from
neurotransmitter
nuclei

signals to
muscles in
face and limbs

signals from
muscles and
joints

signals from
viscera

autonomic signals

*Figure 7-7. Composite of
the diagrams on pages 132,
137, 145, showing the
main body-bound and
brain-bound routes for neural signals involved in
emotion and feeling. Note
that endocrine and other
chemical signals have been
left out for clarity. As in the
previous diagrams, the
basal ganglia have also been
left out.*

evaluative images which caused the body state; and (2) a particular
style and level of efficiency of cognitive process which accompanies
the events described in (1), but is operated in parallel.

The events described in (1) require the enactment of a body state
or of its surrogate within the brain. It presupposes the presence of a
trigger, the existence of acquired dispositions on the basis of which
evaluation will take place, and the existence of innate dispositions
that will activate body-bound responses.

The events described in (2) are triggered from the same system of
dispositions operative in (1), but the target is the set of nuclei in brain
stem and basal forebrain which respond by means of selective neurotransmitter release. The result of the neurotransmitter responses
is a change in the speed at which images are formed, discarded,
attended, evoked, as well as a change in the style of the reasoning
operated on those images. As an example, the cognitive mode which
accompanies a feeling of elation permits the rapid generation of
multiple images such that the associative process is richer and
associations are made to a larger variety of cues available in the

images under scrutiny. The images are not attended for long. The ensuing wealth promotes ease of inference, which may become overinclusive. This cognitive mode is accompanied by an enhancement of motor efficiency and even disinhibition, as well as an increase in appetite and exploratory behaviors. The extreme of this cognitive mode can be found in manic states. By contrast the cognitive mode which accompanies sadness is characterized by slowness of image evocation, poor association in response to fewer clues, narrower and less efficient inferences, overconcentration on the same images, usually those which maintain the negative emotional response. This cognitive state is accompanied by motor inhibition and in general by a reduction in appetite and exploratory behaviors. The extreme of this cognitive mode can be found in depression.[18]

I do not see emotions and feelings as the intangible and vaporous qualities that many presume them to be. Their subject matter is concrete, and they can be related to specific systems in body and brain, no less so than vision or speech. Nor are the responsible brain systems confined to the subcortical sector. Brain core and cerebral cortex work together to construct emotion and feeling, no less so than in vision. One does not see with the cerebral cortex alone, and vision probably begins in the brain stem, in such structures as the colliculi.

Finally it is important to realize that defining emotion and feeling as concrete, cognitively and neurally, does not diminish their loveliness or horror, or their status in poetry or music. Understanding how we see or speak does not debase what is seen or spoken, what is painted or woven into a theatrical line. Understanding the biological mechanisms behind emotions and feelings is perfectly compatible with a romantic view of their value to human beings.

Eight

The Somatic-Marker Hypothesis

REASONING AND DECIDING

WE ALMOST NEVER think of the present, and when we do, it is only to see what light it throws on our plans for the future.[1] These are Pascal's words, and it is easy to see how perceptive he was about the virtual nonexistence of the present, consumed as we are by using the past to plan what-comes-next, a moment away or in the distant future. That all-consuming, ceaseless process of creation is what reasoning and deciding are about, and this chapter is about a fraction of its possible neurobiological underpinnings.

It is perhaps accurate to say that the purpose of reasoning is deciding and that the essence of deciding is selecting a response option, that is, choosing a nonverbal action, a word, a sentence, or some combination thereof, among the many possible at the moment, in connection with a given situation. Reasoning and deciding are so interwoven that they are often used interchangeably. Phillip Johnson-Laird captured the tight interconnection in the form of a saying: "In order to decide, judge; in order to judge, reason; in order to reason, decide (what to reason about)."[2]

The terms reasoning and deciding usually imply that the decider has knowledge (a) about the situation which calls for a decision, (b) about different options of action (responses), and (c) about consequences of each of those options (outcomes) immediately and at future epochs. Knowledge, which exists in memory under dispositional representation form, can be made accessible to consciousness in both nonlanguage and language versions, virtually simultaneously.

The terms reasoning and deciding also usually imply that the decider possesses some logical strategy for producing valid inferences on the basis of which an appropriate response option is selected, and that the support processes required for reasoning are in place. Among the latter, attention and working memory are usually mentioned, but not a whisper is ever heard about emotion or feeling, and next to nothing is ever heard about the mechanism by which a diverse repertoire of options is generated for selection.

From the above accounts of reasoning and deciding, it appears that not all biological processes which culminate in a response selection belong in the scope of reasoning and deciding as outlined above. The following illustrations help make the point.

For the first illustration, consider what happens when the level of your blood sugar drops and neurons in your hypothalamus detect the decline. There is a situation calling for action; there is physiological "know-how" as inscribed in the dispositional representations of the hypothalamus; and, inscribed in a neural circuit, there is a "strategy" to select a response consisting of instituting a hunger state which will eventually drive you to eat. But the process involves no overt knowledge, no explicit display of options and consequences, and no conscious mechanism of inference, up to the point when you become aware of being hungry.

For my second illustration, consider what happens when we move away briskly to avoid a falling object. There is a situation which calls for prompt action (e.g., falling object); there are options for action (to duck or not) and each has a different consequence. However, in order to select the response, we use neither conscious (explicit) knowledge nor a conscious reasoning strategy. The requisite knowledge was once conscious, when we first learned that falling objects may hurt us and that avoiding them or stopping them is better than being hit. But experience with such scenarios as we grew up made our brains solidly pair the provoking stimulus with the most advantageous response. The "strategy" for response selection now consists of activating the strong link between stimulus and response, such that the implementation of the response comes *automatically* and *rapidly*, without effort or deliberation, although one can willfully try to preempt it.

The third illustration pulls together a variety of examples clustered in two groups. One group includes choosing a career; deciding whom to marry or befriend; deciding whether or not to fly when there are impending thunderstorms; deciding whom to vote for or how to invest one's savings; deciding whether to forgive a person who has done you wrong or, if you happen to be a state governor, commute the sentence of the convict now on death row. For most individuals, the other group of examples would also include the reasoning that goes with building a new engine, or designing a building, or solving a mathematical problem, composing a musical piece or writing a book, or judging whether a proposed new law accords with or violates the spirit or letter of a constitutional amendment.

All examples in the third illustration rely on the supposedly clear process of deriving logical consequences from assumed premises, the business of making reliable inferences which, unencumbered by passion, allows us to choose the best possible option, leading to the best possible outcome, given the worst possible problem. It is thus not difficult to separate the third illustration from the former two. In all examples of the third illustration, the stimulus situations have more parts to them; the response options are more numerous; their respective consequences have more ramifications and those consequences are often different, immediately and in the future, thus posing conflicts between possible advantages and disadvantages over varied time frames. Complexity and uncertainty loom so large that reliable predictions are not easy to come by. Just as importantly, a great number of those myriad options and outcomes must appear in consciousness for a management strategy to be engaged. To make a final response selection you must apply reasoning and that involves holding a great many facts in your mind, tallying results of hypothetical actions and matching them against intermediate and ultimate goals, all of which requires a method, some type of game plan among several you rehearsed on countless occasions in the past.

Based on the blatant differences between the third illustration and the former two, it is not surprising to discover that people generally assume that one and the other have entirely unrelated mechanisms, mentally and neurally, so separate indeed that Descartes placed one outside the body, as a hallmark of the human spirit, while the other remained inside, the hallmark of animal spirits; so separate that one stands for clarity of thought, deductive competence, algorithmicity, while the other connotes murkiness and the less disciplined life of the passions.

But if the nature of the examples in the third illustration differs markedly from the first two, it is also true that the examples within it are not all of the same kind. Granted that all require reason in the most common use of the term, some are closer to the person and social environment of the decider than others. Deciding on whom you will love or forgive, making career choices, or choosing an investment are in the immediate personal and social domain; solving Fermat's last theorem or ruling on the constitutionality of a piece of legislation are more removed from the personal core (though one can imagine exceptions). The former align themselves readily with the notions of rationality and practical reason; the latter fall more easily in the general sense of reason, theoretical reason, and even pure reason.

The intriguing notion is that in spite of the manifest differences among the examples and in spite of their apparent clustering by domain and level of complexity, there may well be a common thread running through all of them in the form of a shared neurobiological core.

Reasoning and Deciding in a Personal and Social Space

Reasoning and deciding can be arduous but they are especially so where one's personal life and its immediate social context are concerned. There are good grounds for treating them as a distinctive domain. First, a profound impairment in personal decision-making is not necessarily accompanied by a profound impairment in the nonpersonal domain, as the cases of Phineas Gage, Elliot, and others have confirmed. We are currently investigating how competently can such patients reason when premises do not concern them directly, and how well they can reach the consequent decisions. It may be that the more detached the problems are from their personal and social being, the better they will be at it. Second, common sense observations of human behavior support a similar dissociation in reasoning abilities which cuts in both directions. We all know persons who are exceedingly clever in their social navigation, who have an unerring sense of how to seek advantage for themselves and for their group, but who can be remarkably inept when trusted with a nonpersonal, nonsocial problem. The reverse condition is just as dramatic: We all know creative scientists and artists whose social sense is a disgrace, and who regularly harm themselves and others with their behavior. The absent-minded professor is the benign variety of the latter type. At work, in these different personality styles, are the presence or absence of what Howard Gardner has called "social intelligence," or the presence or absence of one or the other of his multiple intelligences such as the "mathematical."[3]

The personal and immediate social domain is the one closest to our destiny and the one which involves the greatest uncertainty and complexity. Broadly speaking, within that domain, deciding well is selecting a response that will be ultimately advantageous to the organism in terms of its survival, and of the quality of that survival, directly or indirectly. Deciding well also means deciding expeditiously, especially when time is of the essence, and, in the very least, deciding in a time frame deemed appropriate for the problem at hand.

I am aware of the difficulty in defining what is advantageous and I realize that some outcomes may be advantageous for some individuals but not for others. For instance, being a millionaire is not necessarily good, and the same may be true of winning prizes. Much depends on the frame of reference and on the goal we set. Whenever I call a decision advantageous, I refer to basic personal and social outcomes such as survival of the individual and its kin, the securing of shelter, the maintenance of physical and mental health, employment and financial solvency, and good standing in the social group. Gage's or Elliot's new mind no longer permitted them to obtain any of these advantages.

RATIONALITY AT WORK

Let us begin by considering a situation which calls for a choice. Imagine yourself as the owner of a large business, faced with the prospect of meeting or not with a possible client who can bring valuable business but also happens to be the archenemy of your best friend, and proceeding or not with a particular deal. The brain of a normal, intelligent, and educated adult reacts to the situation by rapidly creating scenarios of possible response options *and* related outcomes. To our consciousness, the scenarios are made of multiple imaginary scenes, not really a smooth film, but rather pictorial flashes of key images in those scenes, jump cut from one frame to another, in quick juxtapositions. Examples of what the images would depict include meeting the prospective client; being seen in the client's company by your best friend and placing the friendship in jeopardy; not meeting the client; losing good business but safeguarding the valuable friendship, and so forth. The point I want to stress is that your mind is not a blank at the start of the reasoning process. Rather it is replete with a diverse repertoire of images, generated to the tune of the situation you are facing, entering and exiting your consciousness in a show too rich for you to encompass fully. Even in this caricature you will recognize the sort of quandary we face most every day. How do you resolve the impasse? How do you sort out the questions inherent in the images before your mind's eye?

There are at least two distinct possibilities: the first is drawn from a traditional "high-reason" view of decision making; the second from the "somatic-marker hypothesis."

The "high-reason" view, which is none other than the common-sense view, assumes that when we are at our decision-making best, we are the pride and joy of Plato, Descartes and Kant. Formal logic will, by itself, get us to the best available solution for any problem. An important aspect of the rationalist conception is that to obtain the best results, emotions must be kept *out*. Rational processing must be unencumbered by passion.

Basically, in the high-reason view, you take the different scenarios apart and to use current managerial parlance you perform a cost/benefit analysis of each of them. Keeping in mind "subjective expected utility," which is the thing you want to maximize, you infer logically what is good and what is bad. For instance, you consider the consequences of each option at different points in the projected future and weigh the ensuing losses and gains. Since most problems have far more than the two alternatives in our cartoon, your analysis is anything but easy as you go through your deductions. But notice that even the two-alternative problem is not that simple. Gaining a client may bring immediate reward and also a substantial amount of future reward. How much reward is unknown and so you must estimate its magnitude and rate, over time, so that you can pit it against the potential losses among which you must now count the consequences of losing a friendship. Since the latter loss will vary over time, you must also figure its "depreciation" rate! You are, in fact, faced with a complex calculation, set at diverse imaginary epochs, and burdened with the need to compare results of a different nature which somehow must be translated into a common currency for the comparison to make any sense at all. A substantial part of this calculation will depend on the continued generation of yet more imaginary scenarios, built on visual and auditory patterns, among others, and also on the continued generation of verbal narratives which accompany those scenarios, and which are essential to keep the process of logical inference going.

Now, let me submit that if this strategy is the *only* one you have available, rationality, as described above, is not going to work. At best, your decision will take an inordinately long time, far more than acceptable if you are to get anything else done that day. At worst, you may not even end up with a decision at all because you will get lost in the byways of your calculation. Why? Because it will not be easy to hold in memory the many ledgers of losses and gains that you need to consult for your comparisons. The representations of intermediate steps, which you have put on hold and now need to inspect in order to translate them in whatever symbolic form required to proceed with your logical inferences, are simply going to vanish from your memory slate. You will lose track. Attention and working memory have a limited capacity. In the end, if purely rational calculation is how your mind normally operates, you might choose incorrectly and live to regret the error, or simply give up trying, in frustration.

What the experience with patients such as Elliot suggests is that the cool strategy advocated by Kant, among others, has far more to do with the way patients with prefrontal damage go about deciding than with how normals usually operate. Naturally, even pure reasoners can do better than this with a little help from paper and pencil. Just write down all the options and their myriad unfolding scenarios, and consequences, and so forth. (Apparently that is what Darwin suggested one should do if one wanted to choose the right person to marry.) But first, get a lot of paper and a pencil sharpener, and a large desk, and do not expect anybody to wait until you are finished.

It is also important to note that the flaws of the common-sense view are not confined to the issue of limited memory capacity. Even with paper and pencil to hold the necessary knowledge in place, the reasoning strategies themselves are fraught with weaknesses, as Amos Tversky and Daniel Kahneman have demonstrated.[4] One of those important weaknesses may well be humans' devastating ignorance and defective use of probability theory and statistics, as Stuart Sutherland has suggested.[5] Nonetheless, our brains can often decide well, in seconds, or minutes, depending on the time frame we set as appropriate for the goal we want to achieve, and if they can do so, they must do the marvelous job with more than just pure reason. An alternative view is needed.

THE SOMATIC-MARKER HYPOTHESIS

Consider again the scenarios I outlined. The key components unfold in our minds instantly, sketchily, and virtually simultaneously, too fast for the details to be clearly defined. But now, imagine that *before* you apply any kind of cost/benefit analysis to the premises, and before you reason toward the solution of the problem, something quite important happens: When the bad outcome connected with a given response option comes into mind, however fleetingly, you experience an unpleasant gut feeling. Because the feeling is about the body, I gave the phenomenon the technical term *somatic* state ("soma" is Greek for body); and because it "marks" an image, I called it a *marker*. Note again that I use *somatic* in the most general sense (that which pertains to the body) and I include both visceral and nonvisceral sensation when I refer to somatic markers.

What does the *somatic marker* achieve? It forces attention on the negative outcome to which a given action may lead, and functions as an automated alarm signal which says: Beware of danger ahead if you choose the option which leads to this outcome. The signal may lead you to reject, *immediately*, the negative course of action and thus make you choose among other alternatives. The automated signal protects you against future losses, without further ado, and then allows you *to choose from among fewer alternatives*. There is still room for using a cost/benefit analysis and proper deductive competence, but only *after* the automated step drastically reduces the number of options. Somatic markers may not be sufficient for normal human decision-making since a subsequent process of reasoning and final selection will still take place in many though not all instances. Somatic markers probably increase the accuracy and efficiency of the decision process. Their absence reduces them. This distinction is important and can easily be missed. The hypothesis does not concern the reasoning steps which follow the action of the somatic marker. In short, *somatic markers are a special instance of feelings generated from secondary emotions*. Those emotions and feelings *have been connected, by learning, to predicted future outcomes of certain scenarios*. When a negative somatic marker is juxtaposed to a particular future outcome the combination functions as an alarm bell. When a positive somatic marker is juxtaposed instead, it becomes a beacon of incentive.

This is the essence of the somatic-marker hypothesis. But to get the full scope of the hypothesis you must read on and discover that on occasion somatic markers may operate covertly (without coming to consciousness) and may utilize an "as if" loop.

Somatic markers do not deliberate for us. They assist the deliberation by highlighting some options (either dangerous or favorable), and eliminating them rapidly from subsequent consideration. You may think of it as a system for automated qualification of predictions, which acts, whether you want it or not, to evaluate the extremely diverse scenarios of the anticipated future before you. Think of it as a biasing device. For example, imagine yourself faced with the prospect of an unusually high interest return on an extremely risky investment. Imagine you are asked to say yes or no quickly, in the middle of other distracting business. If a negative somatic state accompanies the thought of proceeding with the investment it will help you reject that option and force a more detailed analysis of its potentially deleterious consequences. The negative state connected with the future counteracts the tempting prospect of an immediate large reward.

The somatic-marker account is thus compatible with the notion that effective personal and social behavior requires individuals to form adequate "theories" of their own minds and of the minds of others. On the basis of those theories we can predict what theories others are forming about our own mind. The detail and accuracy of such predictions is, of course, essential as we approach a critical decision in a social situation. Again, the number of scenarios under scrutiny is immense, and my idea is that somatic markers (or something like them) assist the process of sifting through such a wealth of detail—in effect, reduce the need for sifting because they provide an automated detection of the scenario components which are more likely to be relevant. The partnership between so-called cognitive processes and processes usually called "emotional" should be apparent.

This general account also applies to the choice of actions whose immediate consequences are negative, but which generate positive future outcomes. An example is the enduring of sacrifices *now* in order to attain benefits later. Imagine that in order to turn around the fortunes of your flagging business, you and your workers must accept reduced salaries, starting now, combined with a dramatic increase in the number of work hours. The immediate prospect is unpleasant but the thought of a future advantage creates a positive somatic marker and that overrides the tendency to decide against the immediately painful option. This positive somatic marker which is triggered by the image of a good future outcome must be the base for the enduring of unpleasantness as a preface to potentially better things. How would one otherwise accept surgery, jogging, graduate school, and medical school? By sheer willpower, one might counter, but then how would one explain willpower? Willpower draws on the evaluation of a prospect, and that evaluation may not take place if attention is not properly driven to both the immediate trouble and the future payoff, to both the suffering *now* and the *future* gratification. Remove the latter and you remove the lift from under your willpower's wings. Willpower is just another name for the idea of choosing according to long-term outcomes rather than short-term ones.

An Aside on Altruism

At this stage we may inquire whether the preceding account applies to most if not all decisions which are commonly classified as altruistic, such as the sacrifices that parents make for children, or that just plain good individuals make for other individuals, or that good citizens once made for king and state, and that the remaining heroes of our day still make. In addition to the obvious good that altruists bring to others, they may heap good upon themselves in the form of self-esteem, social recognition, public honor and affection, prestige, and perhaps even money. The prospect of any of those rewards can be accompanied by exaltation (whose neural basis I see as a positive somatic marker), and no doubt can bring even more palpable ecstasy when the prospect becomes reality. Altruistic behaviors benefit their practitioners in yet another way that is relevant here: they save altruists from the future pain and suffering that would have been caused by loss or shame upon *not* behaving altruistically. It is not only that the idea of risking your life to save your child makes you feel good, but that the idea of not saving your child and losing her makes you feel far worse than the immediate risk does. In other words, the evaluation takes place between immediate pain and future reward, *and* between immediate pain and even worse future pain. (A somewhat comparable example is the acceptance of the risks of combat in war. In the past, the social framework in which "moral" wars were waged included a positive payoff for the survivors of combat, and shame and disgrace for those who declined to enter it.)

Does this mean that there is no true altruism? Is this too cynical a view of the human spirit? I do not think so. First, the truth of altruism, or any equivalent behavior, has to do with the relation between what we *internally* believe, feel, or intend, and what we *externally* declare to believe, feel, or intend. Truth does not pertain to the physiological causes that make us believe, feel, or intend in a particular way. Beliefs, feelings, and intentions are indeed the result of a number of factors rooted in our organisms and in the culture in which we have been immersed, even if such factors may be remote and we may not be aware of them. If there are neurophysiological and educational reasons making it likely for some people to be honest and generous, so be it. It does not follow that their honesty and sacrifices are any less meritorious. Moreover, understanding neurobiological mechanisms behind some aspects of cognition and behavior does not diminish the value, beauty, or dignity of that cognition or behavior.

Second, although biology and culture often determine our reasoning, directly or indirectly, and may seem to limit the exercise of individual freedom, we must recognize that humans do have *some* room for such freedom, for willing and performing actions that may go against the apparent grain of biology and culture. Some sublime human achievements come from rejecting what biology or culture propels individuals to do. Such achievements are the affirmation of a new level of being in which one can invent new artifacts and forge more just ways of existing. Under certain circumstances, however, freedom from biological and cultural constraints can also be a hallmark of madness and can nourish the ideas and acts of the insane.

SOMATIC MARKERS: WHERE DO THEY ALL COME FROM?

What is the origin of somatic markers, in neural terms? How have we come to possess such helpful devices? Were we born with them? If not, how did they arise?

As we saw in the previous chapter, we were born with the neural machinery required to generate somatic states in response to certain classes of stimuli, the machinery of primary emotions. Such machinery is inherently biased to process signals concerning personal and social behavior, and it incorporates at the outset dispositions to pair a large number of social situations with adaptive somatic responses. Certain findings in normal humans would fit this view, and so would the evidence for complex patterns of social cognition encountered in other mammals and in birds.[6] Nonetheless, most somatic markers we use for rational decision-making probably were created in our brains during the process of education and socialization, by connecting specific classes of stimuli with specific classes of somatic state. In other words, they are based on the process of secondary emotions.

The buildup of adaptive somatic markers requires that both brain and culture be normal. When *either* brain or culture is defective, at the outset, somatic markers are unlikely to be adaptive. An example of the former can be found at least in some patients affected by a condition known as developmental sociopathy or psychopathy.

Developmental sociopaths or psychopaths are well known to all of us from the daily news. They steal, they rape, they kill, they lie. They are often smart. The threshold at which their emotions kick in, when they do, is so high that they appear unflappable, and are, from their self reports, unfeeling and uncaring. They are the very picture of the cool head we were told to keep in order to do the right thing. In cold blood, and to everybody's obvious disadvantage including their own, sociopaths often repeat their crimes. They are in fact yet another example of a pathological state in which a decline in rationality is accompanied by diminution or absence of feeling. It is certainly possible that developmental sociopathy arises from dysfunction within the same overall system which was impaired in Gage, at cortical or subcortical level. But rather than resulting from blunt macroscopic damage occurring in adulthood, however, the impairment of developmental sociopaths would come from abnormal circuitry and abnormal chemical signaling and begin early in development. Understanding the neurobiology of sociopathy might lead to prevention or treatment. It might also help understand the degree to which social factors interact with biological ones to aggravate the condition, or increase its frequency, and even shed light on conditions which may be superficially similar and yet be largely determined by sociocultural factors.

When the neural machinery that specifically supports the buildup and deployment of somatic markers is damaged in adulthood, as it was in Gage, the somatic-marker device no longer functions properly even if it has been normal until then. I use the term "acquired" sociopathy, as qualified shorthand, to describe a part of the behaviors of such patients although my patients and developmental sociopaths are different in several respects, not the least of which is that my patients are rarely violent.

The effect of a "sick culture" on a normal *adult* system of reasoning seems to be less dramatic than the effect of a focal area of brain damage in that same normal adult system. Yet there are counterexamples. In Germany and the Soviet Union during the 1930s and 1940s, in China during the Cultural Revolution, and in Cambodia during the Pol Pot regime, to mention only the most obvious such cases, a sick culture prevailed upon a presumably normal machinery of reason, with disastrous consequences. I fear that sizable sectors of Western society are gradually becoming other tragic counter-examples.

Somatic markers are thus acquired by experience, under the control of an internal preference system and under the influence of an external set of circumstances which include not only entities and events with which the organism must interact, but also social conventions and ethical rules.

The neural basis for the internal preference system consists of mostly innate regulatory dispositions, posed to ensure survival of the organism. Achieving survival coincides with the ultimate reduction of unpleasant body states and the attaining of homeostatic ones, i.e., functionally balanced biological states. The internal preference system is inherently biased to avoid pain, seek potential pleasure, and is probably pretuned for achieving these goals in social situations.

The external set of circumstances encompasses the entities, physical environment, and events relative to which individuals must act; possible options for action; possible future outcomes for those actions; and the punishment or reward that accompanies a certain option, both immediately and in deferred time, as outcomes of the opted action unfold. Early in development, punishment and reward are delivered not only by the entities themselves, but by parents and other elders and peers, who usually embody the social conventions and ethics of the culture to which the organism belongs. The interaction between an internal preference system and sets of external circumstances extends the repertory of stimuli that will become automatically marked.

The critical, formative set of stimuli to somatic pairings is, no doubt, acquired in childhood and adolescence. But the accrual of somatically marked stimuli ceases only when life ceases, and thus it is appropriate to describe that accrual as a process of continuous learning.

At the neural level, somatic markers depend on learning within a system that can connect certain categories of entity or event with the enactment of a body state, pleasant or unpleasant. Incidentally, it is important not to narrow the meaning of punishment and reward in evolving social interactions. Lack of reward can constitute punishment and be unpleasant, just as lack of punishment can constitute reward and be quite pleasurable. The decisive element is the type of somatic state and feeling produced in a given individual, at a given point in his or her history, in a given situation.

When the choice of option X, which leads to bad outcome Y, is followed by punishment and thus painful body states, the somatic-marker system acquires the hidden, dispositional representation of this experience-driven, noninherited, arbitrary connection. Re-exposure of the organism to option X, or thoughts about outcome Y, will now have the power to reenact the painful body state and thus serve as an automated reminder of bad consequences to come. This is of necessity an oversimplification, but it captures the basic process as I see it. As I will clarify later, somatic markers can operate covertly (they do not need to be perceived consciously) and they can play other helpful roles besides providing signals of "Danger!" or "Go for it!"

A NEURAL NETWORK FOR SOMATIC MARKERS

The critical neural system for the acquisition of somatic-marker signaling is in the prefrontal cortices, where it is in good part coextensive with the system critical for secondary emotions. The neuro-anatomical position of the prefrontal cortices is ideal for the purpose, for the reasons I outline below.

First, the prefrontal cortices receive signals from all the sensory regions in which the images constituting our thoughts are formed, including the somatosensory cortices in which past and current body states are represented continuously. Whether signals arise in perceptions related to the world outside, or in thoughts we are having about the world outside, or in events in the body proper, the prefrontal cortices receive those signals. This is true of all of its separate sectors, because the varied frontal sectors are mutually interconnected within the frontal region itself. The prefrontal cortices thus contain some of the few brain regions to be privy to signals about virtually any activity taking place in our beings' mind or body at any given time.[7] (The prefrontal cortices are not the only eavesdropping posts; another is the entorhinal cortex, the gateway to the hippocampus.)

Second, the prefrontal cortices receive signals from several bioregulatory sectors of the human brain. These include the neurotransmitter nuclei in the brain stem (for instance, those which distribute dopamine, norepinephrine, and serotonin), and in the basal forebrain (those which distribute acetylcholine), as well as the amygdala, the anterior cingulate, and the hypothalmus. One might say of this arrangement that the prefrontal cortices receive messages from the entire staff of the Bureau of Standards and Measures. The innate preferences of the organism related to its survival—its biological value system, so to speak—is conveyed to prefrontal cortices by such signals and is thus part and parcel of the reasoning and decision-making apparatus.

The prefrontal sectors are indeed in a privileged position among other brain systems. Their cortices receive signals about existing and incoming factual knowledge related to the external world; about innate biological regulatory preferences; and about previous and current body state as continuously modified by that knowledge and those preferences. Little wonder that they are so involved with the topic I will address next: the categorization of our life experience according to many contingent dimensions.

Third, the prefrontal cortices themselves represent categorizations of the situations in which the organism has been involved, classifications of the contingencies of our real-life experience. What this means is that prefrontal networks establish dispositional representations for certain combinations of things and events, in one's individual experience, according to the personal relevance of those things and events. Let me explain. In your own life, for example, encounters with a certain type of pleasant but authoritarian person may have been followed by a situation in which you felt diminished or, on the contrary, empowered; being thrust into a leadership role may have brought out the best in you, or the worst; sojourns in the country may have made you melancholic, while the ocean may have made you incurably romantic. Your next-door neighbor may have had precisely the opposite experience, or at least a different one, in each case. This is where the notion of *contingency* applies: it is your own thing, related to your own experience, relative to events that vary with the individual. The experience that you, your neighbor, and I have had with doorknobs or broomsticks might be less contingent, since by and large the structure and operation of that category of entities are consistent and predictable.

Convergence zones located in the prefrontal cortices are thus the repository of dispositional representations for the appropriately categorized and unique contingencies of our life experience. If I ask you to think of weddings, those prefrontal dispositional representations hold the key to such a category and can reconstruct, in your mind's imagetic space, several wedding scenes. (Remember that, neurally speaking, the reconstructions do not occur in prefrontal cortices, but rather in varied early sensory cortices where topographically organized representations can be formed.) If I ask you about Jewish weddings, or Catholic weddings, you might be able to reconstitute the appropriate sets of categorized images and conceptualize one type of wedding or another. Moreover, you might even tell me whether you like weddings, which type you like best, and so forth.

The entire prefrontal region seems dedicated to categorizing contingencies in the perspective of personal relevance. This was first established for the dorsolateral sector, in the work of Brenda Milner, Michael Petrides, and Joaquim Fuster.[8] Work in my laboratory not only supports those observations but suggests that other frontal structures, in the frontal pole and ventromedial sectors, are no less critical for the process of categorization.

Categorized contingencies are the basis for the production of rich scenarios of future outcome required in making predictions and planning. Our reasoning takes into account goals and time scales for the enactment of those goals, and we need a wealth of personally categorized knowledge if we are to preview the unfolding and outcome of scenarios relative to specific goals and in the appropriate time frames.

It is likely that different domains of knowledge are categorized in different prefrontal sectors. Thus the bioregulatory and social domain seem to have an affinity for the systems in the ventromedial sector, while systems in the dorsolateral region appear to align themselves with domains which subsume knowledge of the external world (entities such as objects and people, their actions in space-time; language; mathematics, music).

A fourth reason why the prefrontal cortices are ideally suited for participation in reasoning and deciding is that they are directly connected to every avenue of motor and chemical response available to the brain. The dorsolateral and upper medial sectors can activate the premotor cortices and, from there, bring on-line the so-called primary motor cortex (M1), the supplementary motor area (M2), and the third motor area (M3).[9] The subcortical motor machinery of the basal ganglia is equally accessible to the prefrontal cortices. Last but not least, as first demonstrated by the neuroanatomist Walle Nauta, the ventromedial prefrontal cortices send signals to autonomic nervous system effectors and can promote chemical responses associated with emotion, out of the hypothalamus and brain stem. This demonstration was no coincidence. Nauta was exceptional among neuroscientists in the importance he accorded to visceral information in the cognitive process. In conclusion, the prefrontal cortices and in particular their ventromedial sector are ideally suited to acquire a three-way link among signals concerned with particular types of situations; the different types and magnitudes of body state, which have been associated with certain types of situations in the individual's unique experience; and the effectors of those body states. Upstairs and downstairs come together harmoniously in the ventromedial prefrontal cortices.

SOMATIC MARKERS: THEATER IN THE
BODY OR THEATER IN THE BRAIN?

Given my previous discussion on the physiology of emotions, you should expect not just one mechanism for the somatic-marker process but two. By virtue of the basic mechanism, the body is engaged by the prefrontal cortices and amygdala to assume a particular state profile, whose result is subsequently signaled to the somatosensory cortex, attended, and made conscious. In the alternative mechanism the body is bypassed and the prefrontal cortices and amygdala merely tell the somatosensory cortex to organize itself in the explicit activity pattern that it would have assumed had the body been placed in the desired state and signaled upward accordingly. The somatosensory cortex works as if it were receiving signals about a particular body state, and although the "as if" activity pattern cannot be precisely the same as the activity pattern generated by a real body state, it may still influence decision making.

"As if" mechanisms are a result of development. It is likely that as we were being socially "tuned" in infancy and childhood, most of our decision making was shaped by somatic states related to punishment and reward. But as we matured and repeated situations were categorized, the need to rely on somatic states for every instance of decision making decreased, and yet another level of economic automation developed. Decision-making strategies began depending in part on "symbols" of somatic states. To what extent we depend on such "as if" symbols rather than on the real thing is an important empirical question. I believe this dependence varies widely, from person to person, and from topic to topic. Symbolic processing may be advantageous or pernicious, depending on the topic and the circumstances.

OVERT AND COVERT SOMATIC MARKERS

The somatic marker itself has more than one avenue of action; it has one through consciousness and another outside consciousness.

Whether body states are real or vicarious ("as if"), the corresponding neural pattern can be made conscious and constitute a feeling. However, although many important choices involve feelings, a good number of our daily decisions apparently proceed without feelings. That does not mean that the evaluation that normally leads to a body state has not taken place; or that the body state or its vicarious surrogate has not been engaged; or that the regulatory dispositional machinery underlying the process has not been activated. Quite simply, a signal body state or its surrogate may have been activated but not been made the focus of attention. Without attention, neither will be part of consciousness, although either can be part of a covert action on the mechanisms that govern, without willful control, our appetitive (approach) or aversive (withdrawal) attitudes toward the world. While the hidden machinery underneath has been activated, our consciousness will never know it. Moreover, triggering of activity from neurotransmitter nuclei, which I described as one part of the emotional response, can bias cognitive processes in a covert manner and thus influence the reasoning and decision-making mode.

With due respect for humans and with all the caution that should be associated with comparisons across species, it is apparent that in organisms whose brains do not provide for consciousness and reasoning, covert mechanisms are the core of the decision-making apparatus. They are a means to build "predictions" of outcome and bias the organism's action devices for behaving in a particular way, which may appear to the external observer as a choice. This is, in all likelihood, how worker bumblebees "decide" on which flowers they should land in order to obtain the nectar they need to bring back to the hive. I am not proposing that deep inside each of our brains there is a bee brain deciding for us. Evolution is not the Great Chain of Being, and it has obviously taken many separate roads, one of which led to us. But I believe much can be gained by studying how simpler organisms perform such seemingly complicated tasks with modest neural means. Some mechanisms of the same type may operate in us too. That is all.

Honeysuckle Rose!

"You're confection, goodness knows, honeysuckle rose," so go the naughty lyrics of the Fats Waller jazz standard, and so goes the fate of the busy bee. The reproductive success and ultimate survival of a bee colony depend on how successful the foraging behavior of bumble-bees turns out to be. If they do not work enough at collecting nectar, there will be no honey, and as energy resources dwindle, so will the colony.

Worker bees are equipped with a visual apparatus that allows them to distinguish colors of flowers. They are equipped also with a motor apparatus that allows them to fly and to land. As recent investigations have demonstrated, worker bees learn, after a few visits to flowers of different colors, which are more likely to contain the nectar that they must obtain. It is apparent that, out in a field, they do not land on every possible flower to discover whether there is or not nectar available in each one. They clearly behave as if they predict which flowers are more likely to have nectar, and they land on those flowers more frequently. In the words of Leslie Real, who has experimentally investigated the behavior of worker bumblebees, (*Bombus pennsyl-vanicus*), "Bees appear to form probabilities on the basis of frequency of encounter of different types of reward states, and begin with no prior estimation of likelihoods."[10] How can bees, with their modest nervous systems, produce behavior that is so suggestive of high reason, so seemingly indicative of the use of knowledge, probability theory, and goal-oriented reasoning strategy?

The answer is that the deliberation is apparently achieved by having a simple but powerful system capable of the following: First, detecting stimuli which are innately set as valuable and thus consti-tute a reward; and second, responding to the presence of reward (or lack thereof) with a bias, which can influence the motor system toward a particular behavior (e.g., landing or not), when the situation which delivered (or not) the reward (say, a flower of a given color) appears in the visual field. A recent model has been proposed by Monta ne, Dayan, and Sejnowski for such a system using both behavioral and neurobiological data.[11]

The bee does have a nonspecific neurotransmitter system, which probably uses octopamine, and which is not unlike the dopamine system in mammals. When the reward (nectar) is detected, the nonspecific system can signal to both visual and motor systems and thereby alter their basic behavior. As a result, on the next occasion in which the color that was associated with reward (say, yellow) appears in view, the motor system is prone to land on the flower so colored, and the bee is more likely to find nectar than not. The bee is in fact making a choice, not consciously, not deliberately, but rather using an automated device which incorporates specific natural values, a preference. According to Real, two fundamental aspects of preference must be present: "High expected gain will be preferred to low expected gain; low risk will be preferred to high risk." Incidentally, on the bee's manifestly small memory capacity (it has only short-memory and not an especially large one), the sampling on the basis of which the preference system operates must be extremely small. As few as three visits will apparently do. Again, I am not suggesting at all that our decisions come from a hidden bee brain, but I believe it is important to know that a device as simple as the one outlined above can perform as complicated a task as described here.

INTUITION

Acting at a conscious level, somatic states (or their surrogates) would mark outcomes of responses as positive or negative and thus lead to deliberate avoidance or pursuit of a given response option. But they may also operate covertly, that is, outside consciousness. The explicit imagery related to a negative outcome would be generated, but instead of producing a perceptible body-state change, it would inhibit the regulatory neural circuits located in the brain core, which mediate appetitive, or approach, behaviors. With the inhibition of the tendency to act, or actual enhancement of the tendency to withdraw, the chances of a potentially negative decision would be reduced. In the very least, there would be a gain of time, during which conscious deliberation might increase the probability of making an appropriate (if not the most appropriate) decision. Moreover, a negative option might be voided altogether, or a highly positive one made more likely by enhancement of the impulse to act. This covert mechanism would be the source of what we call intuition, the mysterious mechanism by which we arrive at the solution of a problem *without* reasoning toward it.

The role of intuition in the overall process of making decisions is illuminated in a passage by the mathematician Henri Poincaré, whose insight fits the picture I have in mind:

In fact, what is mathematical creation? It does not consist in making new combinations with mathematical entities already known. Anyone could do that, but the combinations so made would be infinite in number and most of them absolutely without interest. To create consists precisely in not making useless combinations and in making those which are useful and which are only a small minority. Invention is discernment, choice.

How to make this choice, I have before explained; the mathematical facts worthy of being studied are those which, by their analogy with other facts, are capable of leading us to the knowledge of a mathematical law, just as experimental facts lead us to the knowledge of a physical law. They are those which reveal to us unsuspected kinship between other facts, long known, but wrongly believed to be strangers to one another.

Among chosen combinations the most fertile will often be those formed of elements drawn from domains which are far apart. Not that I mean as sufficing for invention the bringing together of objects as disparate as possible; most combinations so formed would be entirely sterile. But certain among them, very rare, are the most fruitful of all.

To invent, I have said, is to choose; but the word is perhaps not wholly exact. It makes one think of a purchaser before whom are displayed a large number of samples, and who examines them, one after the other, to make a choice. Here the samples would be so numerous that a whole lifetime would not suffice to examine them. This is not the actual state of things. The sterile combinations do not even present themselves to the mind of the inventor. Never in the field of his consciousness do combinations appear that are not really useful, except some that he rejects but which have to some extent the characteristics of useful combinations. All goes on as if the inventor were an examiner for the second degree who would only have to question the candidates who had passed a previous examination.[12]

Poincaré's view is similar to the one I am proposing. You do not have to apply reasoning to the entire field of possible options. A preselection is carried out for you, sometimes covertly, sometimes not. A biological mechanism makes the preselection, examines candidates, and allows only a few to present themselves for a final exam. This proposal, it should be noted, is intended cautiously for the personal and social domain for which I have supporting evidence, although Poincaré's insight suggests that the proposal might be extended to other domains.

The physicist and biologist Leo Szilard made a similar point: "The creative scientist has much in common with the artist and the poet. Logical thinking and an analytical ability are necessary attributes to a scientist, but they are far from sufficient for creative work. Those insights in science that have led to a breakthrough were not logically derived from preexisting knowledge: The creative processes on which the progress of science is based operate on the level of the subconscious."[13] Jonas Salk has forcefully articulated the same insight and proposed that creativity rests on a "merging of intuition and reason."[14] It is thus appropriate at this point to say a word about reasoning outside the personal and social realm.

REASONING OUTSIDE THE PERSONAL AND SOCIAL DOMAINS

The squirrel in my backyard that runs up a tree to take cover from the neighbor's adventurous black cat has not reasoned much to decide on his action. He did not really think about his various options and calculate the costs and benefits of each. He saw the cat, was jolted by a body state, and he ran. I am looking at him now, in the solid branch of my pin oak, his heart pounding so strongly that I can see the ribcage flail, his tail beating to the nervous rhythm of squirrel fear. He had a powerful emotion and now he is just upset.

Evolution is thrifty and tinkering. It has had available, in the brains of numerous species, decision-making mechanisms that are body-based and survival-oriented, and those mechanisms have proven successful in a variety of ecological niches. As the environmental contingencies increased and as new decision strategies evolved, it would have made economical sense if the brain structures required to support such new strategies would retain a functional link to their forerunners. Their purpose is the same, survival, and the parameters that control their operation and measure their success are also the same: well-being, absence of pain. Examples abound to demonstrate that natural selection tends to work precisely this way, by conserving something that works, by selecting other devices which can cope with greater complexity, rarely evolving entirely new mechanisms from scratch.

It is plausible that a system geared to produce markers and signposts to guide "personal" and "social" responses would have been co-opted to assist with "other" decision making. The machinery that helps you decide whom to befriend would also help you design a house in which the basement will not flood. Naturally, somatic markers would not need to be perceived as "feelings." But they would still act covertly to highlight, in the form of an attentional mechanism, certain components over others, and to control, in effect, the go, stop, and turn signals necessary for some aspects of decision making and planning in nonpersonal, nonsocial domains. This seems the kind of general marker device that Tim Shallice has proposed for decision making, although he has not specified a neurophysiological mechanism for his markers; in a recent article, Shallice comments on a possible similarity.[15] The underlying physiology might be the same: body-based signaling, conscious or not, on the basis of which attention can be focused.

From an evolutionary perspective, the oldest decision-making device pertains to basic biological regulation; the next, to the personal and social realm; and the most recent, to a collection of abstract-symbolic operations under which we can find artistic and scientific reasoning, utilitarian-engineering reasoning, and the developments of language and mathematics. But although ages of evolution and dedicated neural systems may confer some independence to each of these reasoning/decision-making "modules," I suspect they are all interdependent. When we witness signs of creativity in contemporary humans, we are probably witnessing the integrated operation of sundry combinations of these devices.

THE HELP OF EMOTION, FOR BETTER AND FOR WORSE

The work of Amos Tversky and Daniel Kahneman demonstrates that the objective reasoning we employ in day-to-day decisions is far less effective than it seems and than it ought to be.[16] To put it simply, our reasoning strategies are defective and Stuart Sutherland strikes an important chord when he talks about irrationality as an "enemy within."[17] But even if our reasoning strategies were perfectly tuned, it appears, they would not cope well with the uncertainty and complexity of personal and social problems. The fragile instruments of rationality need special assistance.

The picture is, however, even more complicated than I have suggested so far. Although I believe a body-based mechanism is needed to assist "cool" reason, it is also true that some of those body-based signals can impair the quality of reasoning. Reflecting on the investigations of Kahneman and Tversky, I see some failures of rationality as not just due to a primary calculation weakness, but also due to the influence of biological drives such as obedience, conformity, the desire to preserve self-esteem, which are often manifest as emotions and feelings. For instance, most people fear flying more than they do driving, in spite of the fact that a rational calculation of risk unequivocally demonstrates that we are far more likely to survive a flight between two given cities than a car ride between those two same cities. The difference, by several orders of magnitude, favors flying over driving. And yet most people feel more safe driving than flying. The defective reasoning comes from the so-called "availability error," which, in my perspective, consists of allowing the image of a plane crash, with its emotional drama, to dominate the landscape of our reasoning and to generate a negative bias against the correct choice. The example may appear to be at odds with my main argument but it is not. It shows that biological drives and emotions *can* demonstrably influence decision making, and it suggests that the body-based "negative" influence, although out of step with actual statistics, is nonetheless survival-oriented: planes do crash now and then, and fewer people survive plane crashes than survive car crashes.

But while biological drives and emotion may give rise to irrationality in some circumstances, they are indispensable in others. Biological drives and the automated somatic-marker mechanism that relies on them are essential for some rational behaviors, especially in the personal and social domains, although they can be pernicious to rational decision-making in certain circumstances by creating an overriding bias against objective facts or even by interfering with support mechanisms of decision making such as working memory.

An example from my experience will help clarify the ideas discussed above. Not too long ago, one of our patients with ventromedial prefrontal damage was visiting the laboratory on a cold winter day. Freezing rain had fallen, the roads were icy, and the driving had been hazardous. I had been concerned with the situation and I asked the patient, who had been driving himself, about the ride, about how difficult it had been. His answer was prompt and dispassionate: It had been fine, no different from the usual, except that it had called for some attention to the proper procedures for driving on ice. The patient then went on to outline some of the procedures and to describe how he had seen cars and trucks skidding off the roadway because they were not following these proper, rational procedures. He even had a particular case in point, that of a woman driving ahead of him who had entered a patch of ice, skidded, and rather than gently pulling away from the tailspin, had panicked, hit the brakes, and gone zooming into a ditch. One instant later, apparently unperturbed by this hair-raising scene, my patient crossed the ice patch and drove calmly and surely ahead. He told me all this with the same tranquillity with which he obviously had experienced the incident.

There is not much question that in this instance not having a normal somatic-marker mechanism was enormously advantageous. Most of us would have had to use a deliberate overriding decision to stop us from hitting the brakes, out of panic or out of sheer feeling for the unfortunate driver in front of us. This exemplifies how automated somatic-marker mechanisms can be pernicious to our behavior, and how, under some circumstances, their absence can be an advantage.

The scene now changes to the following day. I was discussing with the same patient when his next visit to the laboratory should take place. I suggested two alternative dates, both in the coming month and just a few days apart from each other. The patient pulled out his appointment book and began consulting the calendar. The behavior that ensued, which was witnessed by several investigators, was remarkable. For the better part of a half-hour, the patient enumerated reasons for and against each of the two dates: previous engagements, proximity to other engagements, possible meteorological conditions, virtually anything that one could reasonably think about concerning a simple date. Just as calmly as he had driven over the ice, and recounted that episode, he was now walking us through a tiresome cost-benefit analysis, an endless outlining and fruitless comparison of options and possible consequences. It took enormous discipline to listen to all of this without pounding on the table and telling him to stop, but we finally did tell him, quietly, that he should come on the second of the alternative dates. His response was equally calm and prompt. He simply said: "That's fine." Back the appointment book went into his pocket, and then he was off.

This behavior is a good example of the limits of pure reason. It is also a good example of the calamitous consequence of not having automated mechanisms of decision making. An automated somatic-marker mechanism would have helped the patient in more ways than one. To begin with, it would have improved the overall framing of the problem. None of us would have spent the amount of time the patient took with this issue, because an automated somatic-marker device would have helped us detect the useless and indulgent nature of the exercise. If nothing else, we would have realized how ridiculous the effort was. At another level, sensing the potentially wasteful approach, we would have opted for one of the alternative dates with the equivalent of tossing a coin or relying on some kind of gut feeling for one or the other date. Or we might simply have turned the decision over to the person asking the question and replied that it really did not matter, that he should choose.

In short, we would picture the waste of time and have it marked as negative; and we would picture the minds of others looking at us, and have that marked as embarrassing. There is reason to believe that the patient did form some of those internal "pictures" but that the absence of a marker prevented those pictures from being properly attended and considered.

If you are wondering how bizarre it is that biological drives and emotion may be *both* beneficial and pernicious, let me say that this would not be the only instance in biology in which a given factor or mechanism may be negative or positive according to the circumstances. We all know that nitric oxide is toxic. It can pollute the air and poison the blood. Yet this same gas functions as a neurotransmitter, sending signals between nerve cells. An even subtler example is glutamate, another neurotransmitter. Glutamate is ubiquitous in the brain, where it is used by one nerve cell to excite another. Yet when nerve cells are damaged, as in a stroke, they release excessive glutamate into the surrounding spaces, and thus cause overexcitation and eventually death of the innocent and healthy nerve cells in the vicinity.

Ultimately, the question raised here concerns the type and amount of somatic marking applied to different frames of the problem being solved. The airline pilot in charge of landing his aircraft in bad weather at a busy airport must not allow feelings to perturb attention to the details on which his decisions depend. And yet he must have feelings to hold in place the larger goals of his behavior in that particular situation, feelings connected with the sense of responsibility for the life of his passengers and crew, and for his own life and that of his family. Too much feeling at the smaller frames or too little at the larger frame can have disastrous consequences. Floor traders at a stock exchange are in a similar predicament.

A fascinating illustration of these points can be found in a study involving Herbert von Karajan.[18] The Austrian psychologists G. and H. Harrer were allowed to observe the pattern of von Karajan's autonomic responses in several circumstances: while he landed his private jet at the Salzburg airport, while he conducted in the recording studio, and while he listened to the playback of the recorded piece (the piece was Beethoven's *Leonora* Overture No. 3).

Von Karajan's musical performance was punctuated by large response changes. His pulse rate went up more dramatically during passages of emotional impact than during passages of actual physical exertion. The profile of his pulse rate when he listened to the playback was parallel to that obtained during the recording. The good news is that Mr. Karajan landed his plane like a dream and even when he was told, after touchdown, to make an emergency takeoff in a steep ascent angle, his pulse increased a bit but far less so than during his musical exercises. His heart was in the music, as well it should have been, and as I once discovered personally at a concert: Just before he lowered the baton to begin a performance of Beethoven's Sixth, I whispered something to my wife, who was sitting next to me. Von Karajan froze the movement of his arm, turned around, and fulminated at me with his eyes. Too bad nobody measured our respective pulses.

BESIDE AND BEYOND SOMATIC MARKERS

Necessary as something like the somatic-marker mechanism may be to construct a neurobiology of rationality, it is apparent that necessity does not make for sufficiency. As I indicated in my account, logical competence does come into play beyond somatic markers. Moreover, several processes must precede, co-occur with, or immediately follow somatic markers, to permit their operation. What are those processes, and can anything be ventured about their neural substrate?

What else happens when somatic markers, overtly or covertly, do their biasing job? What happens in your brain so that the images over which you reason are sustained over the necessary time intervals? To address these questions, let us return to a problem outlined at the beginning of the chapter. What dominates the mind landscape once you are faced with a decision is the rich, broad display of knowledge about the situation that is being generated by its consideration. Images corresponding to myriad options for action and myriad possible outcomes are activated and keep being brought into focus. The language counterpart of those entities and scenes, the words and sentences that narrate what your mind sees and hears, is there too, vying for the spotlight. This process is based on a continuous creation of combinations of entities and events, resulting in a richly diverse juxtaposition of images which accords with previously categorized knowledge. Jean-Pierre Changeux has proposed the descriptor "generator of diversity" for the prefrontal structures which presumably carry out this function and lead to the formation of a large repertoire of images elsewhere in the brain. This is an especially apt descriptor since it conjures up its immunological forerunner, and generates itself a curious acronym.[19]

This generator of diversity requires a vast store of factual knowledge, about the situations we may face, about the actors in those situations, about what they can do and how their varied actions produce varied outcomes. Factual knowledge is categorized (the facts that constitute it being organized by classes, according to constituent criteria), and categorization contributes to decision making by classifying types of options, types of outcomes, and connections of options to outcomes. Categorization also ranks options and outcomes relative to some particular value. When we face a situation, prior categorization allows us to discover rapidly whether a given option or outcome is likely to be advantageous, or how diverse contingencies can modify the degree of advantage.

The process of knowledge display is possible only if two conditions are met. First, one must be able to draw on mechanisms of *basic attention,* which permit the maintenance of a mental image in consciousness to the relative exclusion of others. In neural terms, this probably depends on enhancement of the neural activity pattern that sustains a given image, while other neural activity around it is depressed.[20] Second, one must have a mechanism of *basic working memory,* which holds separate images for a relatively "extended" period of hundreds to thousands of milliseconds (from tenths of a second to a number of consecutive seconds).[21] This means that the brain reiterates over time the topographically organized representations supporting those separate images. There is, of course, an important question to be asked at this point: what drives basic attention and working memory? The answer can only be *basic value,* the collection of basic preferences inherent in biological regulation.

Without basic attention and working memory there is no prospect of coherent mental activity, and, to be sure, somatic markers cannot operate at all, because there is no stable playing field for somatic markers to do their job. However, attention and working memory probably continue to be required even after the somatic-marker mechanism operates. They are necessary for the process of reasoning, during which possible outcomes are compared, rankings of results are established, and inferences are made. In the full somatic-marker hypothesis, I propose that a somatic state, negative or positive, caused by the appearance of a given representation, operates not only as a *marker for the value of what is represented, but also as a booster for continued working memory and attention.* The proceedings are "energized" by signs that the process is actually being evaluated, positively or negatively, in terms of the individual's preferences and goals. The allocation and maintenance of attention and working memory do not happen by miracle. They are first motivated by preferences inherent in the organism, and then by preferences and goals acquired on the basis of the inherent ones.

In terms of the prefrontal cortices, I am suggesting that somatic markers, which operate on the bioregulatory and social domain aligned with the ventromedial sector, influence the operation of attention and working memory within the dorsolateral sector, the sector on which operations on other domains of knowledge depend. This leaves open the possibility that somatic markers also influence attention and working memory within the bioregulatory and social domain itself. In other words, in normal individuals, somatic markers which arise out of activating a particular contingency boost attention and working memory throughout the cognitive system. In patients with damage in the ventromedial region, all of these actions would be compromised to a smaller or greater degree.

BIASES AND THE CREATION OF ORDER

There are thus three supporting players in the process of reasoning over a vast landscape of scenarios generated from factual knowledge: *automated somatic states*, with their biasing mechanisms; *working memory*; and *attention*. All three supporting players interact and all three seem concerned with the critical problem of creating order out of parallel spatial displays, a problem first recognized by Karl Lashley, which arises because the brain's design only permits, at any one time, a limited amount of conscious mental output and movement output.[22] The images which constitute our thoughts must be structured in "phrases," which in turn must be "sententially" ordered in time, just as the frames of movement which constitute our external responses must be "phrased" in a particular way and those phrases placed in a particular "sentential" order for a motion to have its desired effect. The selection of the frames that end up composing the "phrases" and "sentences" of our mind and movement is made from a parallel display of possibilities. And because both thought and movement require concurrent processing, the organization of several ordered sequences must go on continuously.

Whether we conceive of reason as based on automated selection, or on a logical deduction mediated by a symbolic system, or— preferably—both, we cannot ignore the problem of order. I propose the following solution: (1) If order is to be created among available possibilities, then they must be ranked. (2) If they are to be ranked, then criteria are needed (values or preferences are equivalent terms). (3) Criteria are provided by somatic markers, which express, at any given time, the cumulative preferences we have both received and acquired.

But how do somatic markers function as criteria? One possibility is that when different somatic markers are juxtaposed to different combinations of images, they modify the way the brain handles them, and thus operate as a bias. The bias might allocate attentional enhancement differently to each component, the consequence being the automated assigning of *varied degrees* of attention to *varied contents*, which translates into an uneven landscape. The focus of conscious processing could be driven then from component to component, for instance, according to their rank in a progression. For all this to happen, the components must remain displayed for an interval of time of hundreds to a few thousand milliseconds, in relatively stable fashion, and that is what working memory achieves. (I found some support for this general idea in recent studies on the neurophysiology of perceptual decision by William T. Newsome and his colleagues. A change in the balance of signals applied to a particular neuron population representing a particular content resulted in a "decision" in favor of that content by what appeared to be a "winner-take-all" mechanism.[23])

Normal cognition and movement require organization of concurrent and interactive sequences. Where there is a need for order there is a need for decision, and where there is a need for decision there must be a criterion to make that decision. Since many decisions have an impact on an organism's future, it is plausible that some criteria are rooted, directly or indirectly, in the organism's biological drives (its reasons, so to speak). Biological drives can be expressed overtly and covertly, and used as a marker bias enacted by attention in a field of representations held active by working memory.

The automated somatic-marker device of most of us lucky enough to have been reared in a relatively healthy culture has been accommodated by education to the standards of rationality of that culture. In spite of its roots in biological regulation, the device has been tuned to cultural prescriptions designed to ensure survival in a particular society. If we assume that the brain is normal and the culture in which it develops is healthy, the device has been made rational relative to social conventions and ethics.

The action of biological drives, body states, and emotions may be an indispensable foundation for rationality. The lower levels in the neural edifice of reason are the same that regulate the processing of emotions and feelings, along with global functions of the body proper such that the organism can survive. These lower levels maintain direct and mutual relationships with the body proper, thus placing the body within the chain of operations that permit the highest reaches of reason and creativity. Rationality is probably shaped and modulated by body signals, even as it performs the most sublime distinctions and acts accordingly.

David Hume, who was keenly aware of the value of the emotions, might not disagree with the statements above, and Pascal, who said that "the heart has reasons that reason does not know at all," might have found the preceding account plausible.[24] If I might be permitted to modify his statement: *The organism has some reasons that reason must utilize.* That the process continues beyond the reasons of the heart is not in doubt. For one thing, using the instruments of logic, we can check on the validity of the selections our preferences have helped make. For another, we can go beyond them using the strategies of deduction and induction in readily available language propositions. (After completing this manuscript, I came across several compatible voices. J. St. B.T. Evans has recently proposed that there are two types of rationality, largely concerned with the two domains I have outlined here [personal/ social and not]; the philosopher Ronald De Sousa has argued that emotions are inherently rational; and P.N. Johnson-Laird and Keith Oatley have suggested that basic emotions help manage actions in a rational way.[25])

Part

3

Nine

Testing the Somatic-
Marker Hypothesis

TO KNOW BUT NOT TO FEEL

MY FIRST APPROACH in investigating the somatic-marker hypothesis involved the use of autonomic nervous system responses, in a series of studies I undertook with Daniel Tranel, a psychophysiologist and experimental neuropsychologist. The autonomic nervous system consists of both autonomic control centers, located within the limbic system and brain stem (the amygdala being the prime example), and neuron projections arising from those centers and aimed at viscera throughout the organism. Blood vessels everywhere, including those in the thick of the most extensive organ in the body, the skin, are innervated by terminals from the autonomic nervous system, and so are the heart, the lung, the gut, the bladder, and the reproductive organs. Even an organ such as the spleen, which is concerned largely with immunity, is innervated by the autonomic nervous system.

The autonomic nerve branches are organized in two large divisions, the sympathetic and the parasympathetic, and they travel from the brain stem and the spinal cord, sometimes on their own, sometimes accompanying nonautonomic nerve branches. (The actions of the sympathetic and parasympathetic divisions are mediated by different neurotransmitters and are largely antagonic, e.g., where one promotes contraction of smooth muscle, the other promotes dilation.) The returning autonomic nerve branches, which bring signals concerning the state of the viscera to the central nervous system, tend to use the same routes.

From the point of view of evolution, it appears that the autonomic nervous system was the neural means by which the brain of organisms far less sophisticated than we are, intervened in the regulation of their internal economy. When life consisted mainly of securing the balanced function of a few organs, and when there was a limited type and number of transactions with the surrounding environment, the immune and endocrine systems governed most of what there was to govern. What the brain required was some signal about the state of varied organs, along with a means to modify that state given a particular external circumstance. The autonomic nervous system provided precisely that: an incoming network for signaling changes in viscera, and an outgoing network for motor commands to those viscera. Later, there evolved more complex forms of motor response, such as those which eventually controlled the hands and the vocal apparatus. The latter responses required a progressively more complex differentiation of the peripheral motor system so that it could control fine muscle and joint operations, as well as signal touch, temperature, pain, the position of joints, and the degree of muscle contraction.

Recall that the idea of the somatic marker encompasses an integral change of body state, which includes modifications in both the viscera and the musculoskeletal system, induced by both neural signals and chemical signals, although the visceral component seems somewhat more critical than the musculoskeletal in the construction of background and emotional states. In order to begin exploring the somatic-marker hypothesis experimentally, we had to choose some aspect of this vast panorama of changes, and it made sense to start by studying autonomic nervous system responses. After all, when we generate the somatic state that characterizes a certain emotion, the autonomic nervous system is probably the key to achieving the appropriate modification of physiological parameters in the body, notwithstanding the important chemical routes that are activated at the same time.

Among the autonomic nervous system responses that can be investigated in the laboratory, the skin conductance response is perhaps the most useful. It is easy to elicit, it is reliable, and it has been studied thoroughly by psychophysiologists, in normal individuals of various ages and cultures. (Many other responses, such as heart rate and skin temperature, have also been studied.) The skin conductance response can be recorded, without any pain or discomfort to the subject, by using a pair of electrodes connected to the skin and a polygraph. The principle behind the response is as follows: As our body begins to change after a given percept or thought, and as a related somatic state begins to be enacted (for instance, that of a given emotion), the autonomic nervous system subtly increases the secretion of fluid in the skin's sweat glands. Although the increase in quantity of fluid is usually so small that it is not noticeable to the naked eye or to the neural sensors in one's own skin, it is sufficient to reduce resistance to the passage of an electrical current. To measure the response, then, the experimenter passes a low-voltage electrical current in the skin between two detector electrodes. The skin conductance response consists of a change in the amount of current conducted. The response is recorded as a wave, which takes time to rise and then fall. The amplitude of the wave can be measured (in microSiemens), as can its profile in time; the frequency with which responses occur relative to a particular stimulus, over any specified time interval, can also be measured.

Skin conductance responses have been a staple of investigative psychophysiology, and they have had a practical and often controversial role in so-called lie-detector tests, whose purpose obviously differs from that of our experiments. These tests aim at determining if subjects are lying, by tricking them into denying knowledge of a particular object or person which makes them unwittingly produce a skin conductance response.

In our study, we wanted to determine first of all whether patients such as Elliot could still generate skin conductance responses. Was their brain still capable of triggering a change in somatic state at all? To answer this question, we compared patients who had frontal lobe damage with normal individuals and with patients who had damage elsewhere in the brain, in experimental conditions known to elicit a skin conductance response consistently, and thus indicate the normalcy of the neural machinery used for skin conductance responses. One such condition is known as "startle," and consists of surprising the subject with an unexpected sound, for instance the clapping of hands, or with the unexpected glare of light caused by a strobe lamp flickering rapidly. Another reliable indicator of normalcy in the skin conductance machinery is a simple physiological act, such as taking a deep breath.

It did not take long for us to verify that all of our subjects with frontal lobe damage could elicit skin conductance responses under the experimental conditions just as well as did normals and patients without frontal lobe damage. In other words, in the patients with frontal damage nothing essential seemed to have been disturbed in the neural machinery with which skin conductance responses are elicited.

We wondered whether patients with frontal lobe damage would generate skin conductance responses to a stimulus that required an evaluation of its emotional content. Why was this a relevant question? Because patients such as Elliot had an impairment in their experience of emotion, and because we knew, from previous studies in normals, that when we are exposed to stimuli with a high emotional content, they reliably produce strong skin conductance responses. We generate such responses when we view scenes of horror or physical pain, or photographs of such scenes, or when we view sexually explicit images. You can imagine the skin conductance response as the subtle, imperceptible part of a body state that, if it unfolds completely, will give you the perceptible sense of excitement and arousal—goose pimples, in some people. But it is important to realize that because skin conductance changes are only a part of the body-state response, having these changes does not guarantee that you will end up perceiving a notable body-state change. This, though, seems true: If you do not have a skin conductance response, it does not appear that you ever will have the conscious body state characteristic of an emotion.

We set up the experiment such that we could compare patients with frontal damage with both normal individuals and patients without frontal damage, making sure that all subjects had been matched for age and educational level. The subjects were to view a succession of projected slides while sitting comfortably in a chair, hooked to a polygraph, saying nothing and doing nothing. Many of the slides were perfectly banal, showing bland scenery or abstract patterns, but every now and then, randomly, a slide with a disturbing image would appear. The experiment ran for as long as there were slides to view, and there were hundreds of them. The subjects had been told before the projection began that they should be attentive, since later, during a debriefing period, they would be asked to tell us about what they saw, how they felt about it, and even *when* they saw given pictures relative to the entire period of the experiment.

The results were unequivocal.[1] The subjects without frontal damage—both the normal individuals and those with brain damage which did not affect the frontal lobes—generated abundant skin conductance responses to the disturbing pictures but not to the bland ones. On the contrary, the patients with frontal lobe damage failed to generate *any* skin conductance responses whatsoever. Their recordings were flat. (See Figure 9-1.)

Before jumping to conclusions we decided to repeat the experiment with different pictures and different subjects, and to repeat the experiment with the same subjects at a different time. These manipulations did not change the results. Again and again, under the passive conditions described above, it was the frontally damaged subjects who did not generate any skin conductance response to the disturbing images, even though afterward they could discuss the

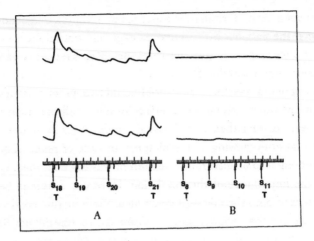

Figure 9-1. The profile of skin conductance responses in normal controls without brain damage (A) and in patients with frontal lobe damage (B), when they viewed a sequence of pictures, some of which had a strong emotional content (identified by a T, for "target," under the stimulus number, e.g., $S_{18}T$), and some of which did not. Normal controls produce large responses shortly after viewing "emotional" images but not after neutral ones. Frontal patients do not respond to either.

content of those slides in detail and even recall the position in time at which certain slides had appeared in the set. They were able to describe, in words, the fear, disgust, or sadness of the pictures they saw, and they were able to tell us how recently a particular picture had been seen relative to another, or how early or late one had appeared in the entire set. There was no question that these subjects had been attentive to the slide show, that they had understood the content of the images, and that the concepts represented in them were available to the subjects on various levels—they knew not only what they depicted (e.g., that there had been a homicide) but they also knew that the way in which the homicide was represented had an element of horror, or that one should be sorry for the victim and regret that such a situation had come to pass. In other words, a given stimulus had produced an abundant evocation of knowledge perti- nent to the situation represented in the stimulus in the mind of the frontal subjects performing the experiment. Yet, unlike the control subjects, the patients with frontal damage had not elicited a skin conductance response. The analysis of the differences revealed that they were highly significant.

During one of the very first debriefing interviews, one particular patient, spontaneously and with perfect insight, confirmed to us that more was missing than just the skin conductance response. He noted that after viewing all the pictures, in spite of realizing their content ought to be disturbing, he himself was not disturbed. Consider the importance of this revelation. Here was a human being cognizant of both the manifest meaning of these pictures and their implied emotional significance, but aware also that he did not "feel" as he knew he used to feel—and as he was perhaps "supposed" to feel?—relative to such implied meaning. The patient was telling us, quite plainly, that his flesh no longer responded to these themes as it once had. That somehow, *to know does not necessarily mean to feel*, even when you realize that what you know ought to make you feel in a specific way but fails to do so.

The consistent lack of skin conductance responses, together with the testimony of frontally damaged patients about the absence of feeling, convinced us, more than any other result, that the somatic-marker hypothesis was worth pursuing. It seemed, indeed, as if those patients' entire scope of knowledge was available except for the dispositional knowledge pairing a particular fact with the mechanism to reenact an emotional response. In the absence of that automated link, the patients could evoke factual knowledge internally but could not produce a somatic state or, in the very least, a somatic state of which they could be aware. They could avail themselves of abundant factual knowledge but could not experience a feeling, that is, the "knowledge" of how their bodies ought to behave relative to the evoked factual knowledge. And because these individuals had previously been normal, they were able to realize that their comprehensive mental state was not as it should have been, that something was lacking.

As a whole, the skin conductance response experiments gave us a measurable physiological counterpart to the observable reduction in emotional resonance we had noted in these patients, and to their own perceived reduction in feeling.

RISK TAKING: THE GAMBLING EXPERIMENTS

Another approach we took to testing the somatic-marker hypothesis made use of a task designed by my postdoctoral student Antoine Bechara. Frustrated, as all researchers are, by the artificial nature of most experimental neuropsychological tasks, he wanted to develop as lifelike a means as possible to assess decision-making performance. The clever set of tasks that he devised, and further refined in collaboration with Hanna Damasio and Steven Anderson, have come to be known in our laboratory, predictably enough, as the "Gambling Experiments."[2] Overall, the setting for the experiments is colorful, a far cry from the boring manipulations of most other such situations. Normals and patients alike enjoy it, and the nature of the investigation makes for amusing episodes. I recall the bulging eyes and dropped jaw of a distinguished visitor who came to my office after walking by the lab where an experiment was in progress. "There are people *gambling* here!" he informed me in a whisper.

In the basic experiment, the subject, known as the "Player," sits in front of four decks of cards labeled A, B, C, and D. The Player is given a loan of $2,000 (play money but looking like the real thing) and told that the goal of the game he is about to play is to lose as little as possible of the loan and try to make as much extra money as possible. Play consists of turning cards, one at a time, from any of the four decks, until the experimenter says to stop. The Player thus does not know the total number of turns required to end the game. The Player is told also that turning any and every card will result in earning a sum of money, and that every now and then turning some cards will result in both earning money and having to pay a sum of money to the experimenter. Neither the amounts of gain or loss in any card, nor the cards' connection to a specific deck, nor the order of their appearance is disclosed at the outset. The amount to be earned or paid with a given card is disclosed only after the card is turned. No other instruction is provided. The tally of how much has been earned or lost at any point is not disclosed, and the subject is not allowed to keep written notes.

The turning of any card in decks A and B pays a handsome $100, while the turning of any card in decks C and D only pays $50. Cards keep being turned on any deck, and quite unpredictably, certain cards in decks A and B (the $100-paying decks) require the Player to make a sudden high payment, sometimes as much as $1,250. Likewise, certain cards in decks C and D (the $50-paying decks) also require a payment, but the sums are much smaller, less than $100 on the average. These undisclosed rules are never changed. Unbeknownst to the Player, the game will be terminated after 100 plays. There is no way for the Player to predict, at the outset, what will happen, and no way to keep in mind a precise tally of gains and losses as the game proceeds. Just as in life, where much of the knowledge by which we live and by which we construct our adaptive future is doled out bit by bit, as experience accrues, uncertainty reigns. Our knowledge—and the Player's—is shaped by both the world with which we interact and by the biases inherent in our organism, for example, our preferences for gain over loss, for reward over punishment, for low risk over high risk.

What regular folks do in the experiment is interesting. They begin by sampling from all four decks, in search of patterns and clues. Then, more often than not, perhaps lured by the experience of high reward from turning cards in the A and B decks, they show an early preference for those decks. Gradually, however, within the first thirty moves, they switch the preference to decks C and D. In general, they stick to this strategy until the end, although self-professed high-risk players may resample decks A and B occasionally, only to return to the apparently more prudent course of action.

There is no way for players to carry out a precise calculation of gains and losses. Rather, bit by bit, they develop a hunch that some decks—namely, A and B—are more "dangerous" than others. One might say they intuit that the lower penalties in decks C and D will make them come out ahead in the long run, despite the smaller initial gain. I suspect that before and beneath the conscious hunch there is a nonconscious process gradually formulating a prediction for the outcome of each move, and gradually telling the mindful player, at first softly but then ever louder, that punishment or reward is about to strike *if* a certain move is indeed carried out. In short, I doubt that it is a matter of only fully conscious process, or only fully nonconscious process. It seems to take both types of processing for the well-tempered decision-making brain to operate.

The behavior of ventromedial frontal patients in this experiment was most informative. What they did in the card game resembled what they often have done in real life since they sustained their brain lesion, and differed from what they would have done before the lesion. Their behavior was diametrically opposed to that of normal individuals.

After an early general sampling, the frontally damaged patients systematically turned more cards in the A and B decks, and fewer and fewer cards in the C and D decks. Despite the higher amount of money they received from turning the A and B cards, the penalties they kept having to pay were so high that halfway through the game they were bankrupt and needed to make extra loans from the experimenter. In the case of Elliot, who played the game, this behavior is especially remarkable because he still describes himself as a conservative, low-risk person, and because even normal subjects who described themselves as high-risk and as gamblers performed so differently, and so prudently. Moreover, at the end of the game, Elliot knew which decks were bad and which were not. When the experiment was repeated a few months later, with different cards and different labels for the decks, Elliot behaved no differently from how he did in real-life situations, where his errors have persisted.

This is the first laboratory task in which a counterpart to Phineas Gage's troubled real-life choices has been measured. Patients with frontal lobe lesions whose behavior and lesions are comparable to Elliot's have performed with a pattern similar to his, in this task.

Why should this task succeed where others fail? Probably because it mimics life so closely. The task is carried out in real time and resembles regular card games. It factors in punishment and reward, and overtly includes monetary values. It engages the subject in a quest for advantage, it poses risks, and it offers choices but does not tell how, when, or what to choose. It is full of uncertainty, and the only way to minimize that uncertainty is to generate hunches, estimates of probability, by whatever means possible, since precise calculation is not possible.

The neuropsychological mechanisms behind this behavior are fascinating, in particular for the frontally damaged patients. Clearly Elliot was engaged in the task, fully attentive, cooperative, and interested in the outcome. In fact, he wanted to *win*. What made him choose so disastrously? As with his other behaviors, we can invoke neither lack of knowledge nor lack of understanding of the situation. As the game progressed, the premises for the choices were constantly available. When he lost $1,000, he realized it, since he paid

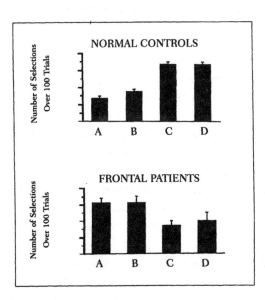

Figure 9-2. A bar graph with results of gambling task relative to each deck. Normal controls prefer decks C and D overall, while frontal patients do the opposite. The differences are significant.

the penalty to the observer. And yet he persisted in choosing the $100-paying decks, which brought him loss every time he was penalized. We cannot even suggest that a continuation of the game required an added memory load, because the continued dire or positive results were made explicit, so often. As their losses accumulated, Elliot and the other frontally damaged patients had to take loans which served as obvious proof of the negative course of their playing. And yet they persisted in making the least advantageous choices for longer than any other group of subjects so far observed in this task, including several patients with brain damage outside the frontal lobes.

Patients with large lesions elsewhere in the brain—for instance, outside the prefrontal sectors—can play the gambling game as normals do provided they can see and can understand the instructions. This is even true of patients with language impairment. A patient with a severe naming defect caused by dysfunction of the left temporal cortex played the entire game worrying aloud, in her broken, aphasic language, that she could not make any sense of what was going on. Yet her performance profile was flawless. She unflinchingly chose what her perfectly intact rationality led her to choose.

What could have been happening in the brains of the frontally damaged subjects? A list of some possible alternative mechanisms ran as follows:

1. They are no longer sensitive to punishment as normal subjects are, and are controlled only by reward.
2. They have become so sensitive to reward that its mere presence makes them overlook punishment.
3. They are still sensitive to punishment and reward but neither punishment nor reward contributes to the automated marking or maintained deployment of predictions of future outcomes, and as a result immediately rewarding options are favored.

In trying to sort out among these possibilities, Antoine Bechara developed another task that consisted of inverting the schedules of reward and punishment. Now punishment came first, in the form of large or not-so-large payments with every card-turning, while reward came interspersed with the turning of some cards. As was the case in the first game, two decks yielded a gain and two decks yielded a loss. In this new task Elliot performed pretty much as normal subjects, and the same was true of other frontal lobe patients. In other words, the idea that Elliot and other frontally damaged patients were merely insensitive to punishment could not be correct.

Another bit of evidence we adduced against the hypothesis of insensitivity to punishment came from a qualitative analysis of the patients' performance in the first task. The profiles showed that immediately after making a penalty payment, the patients avoided the deck from which the bad card had come, just as normal subjects did, but then, unlike normals, they returned to the bad deck. This also suggests that the patients were still sensitive to punishment, although the effects of punishment did not seem to last for very long, probably because it was not connected with the formulation of predictors concerning future prospects.

MYOPIA FOR THE FUTURE

To an external observer, the mechanisms outlined in the third hypothesis would make patients seem far more concerned with the present than with the future. Deprived of the marking or sustained deployment of predictions of the future, these patients are controlled largely by immediate prospects and indeed appear insensitive to the future. This suggests that patients with frontal lobe damage suffer from a profound exaggeration of what may be a normal basic tendency, to go for the now rather than bank on the future. But whereas the tendency is brought under control in normal and socially adapted individuals, especially in situations where it does matter personally, the magnitude of the tendency becomes so overwhelming in frontal lobe patients that they easily succumb. We might describe the predicament of these patients as a "myopia for the future," a concept that has been proposed to explain the behavior of individuals under the influence of alcohol and other drugs. Inebriation does narrow the panorama of our future, so much so that almost nothing but the present is processed with clarity.[3]

We might conclude that the result of these patients' lesions is the discarding of what their brains have acquired through education and socialization. One of the most distinctive human traits is the ability to learn to be guided by future prospects rather than by immediate outcomes, something we begin to acquire in childhood. In frontal lobe patients, brain damage not only compromises the repository of knowledge pertinent to such guidance that had been accumulated until then, but further compromises the ability to acquire new knowledge of the same type. The only redeeming aspect of this tragedy, as is often the case in instances of brain damage, resides with the window it opens for science. Some insight can indeed be gained into the nature of the processes that have been lost.

We know where the lesions that cause the problem are. We know something about the neural systems contained in the areas damaged by those lesions. But why is it that their destruction suddenly makes future consequences no longer have an impact in decision making? When we analyze the process into its components, we come up with various possibilities.

It is conceivable that the images which constitute a future scenario are weak and unstable. The images would be activated but somehow not held long enough in consciousness to play a role in the appropriate reasoning strategy. In neuropsychological terms this is equivalent to saying that working memory and/or attention are not functioning well, as far as images about the future are concerned. This account works regardless of whether the images concern the domain of body states or the domain of facts external to the body.

Another account uses the idea of somatic markers. Even if the images of future consequences were stable, damage in the ven-

tromedial prefrontal cortices would preclude the evocation of pertinent somatic-state signals (through either a body loop or an "as if"
loop), and consequently the relevant future scenarios would no
longer be marked. Their significance would not be apparent, and
their impact on the decision-making process would be voided, or
easily overcome by the significance of immediate prospects. I can
unpack this account a bit further by saying that what would be lost is
a mechanism to generate automated predictions of the significance
of a future outcome. In normal subjects participating in the gambling experiments described above, the significance would have been
acquired from repeated exposure to different ratios of punishment
and reward relative to a given deck. In other words, the brain would
associate a certain degree of badness and goodness with each deck,
A, B, C, and D. The basic process would be nonconscious and would
consist of a weighing of frequency and amount of negative states.
The neural expression of this covert, nonconscious means of reasoning would be the biasing somatic state. No such process seems to
happen in frontally damaged patients.

My current view combines the two possibilities. Activation of
pertinent somatic states is the critical factor. But I also suspect that
the somatic-state mechanism acts as a booster to maintain and
optimize working memory and attention concerned with scenarios of
the future. In short, you cannot formulate and use adequate "theories" for your mind and for the mind of others if something like the
somatic marker fails you.

PREDICTING THE FUTURE:
PHYSIOLOGICAL CORRELATES

A natural follow-up to the gambling experiments was suggested by
Hanna Damasio. Her idea was to monitor the performance of both
normal subjects and those who had frontal damage, with skin conductance responses during gambling tasks. In what ways would the
patients behave differently from normals?

Antoine Bechara and Daniel Tranel set out to investigate this question by having patients and normal subjects play the card game while hooked to the polygraph. Two sets of parallel data were thus collected: the continuous choices the subjects were making as they went along, and the continuous profile of skin conductance responses generated in the process.

The first batch of results yielded a striking profile. Both normal controls and frontal lobe patients generated skin conductance responses as each reward or punishment occurred after turning an appropriate card. In other words, within the few seconds immediately following their receiving the monetary reward or having to pay the penalty, normal subjects as well as frontally damaged subjects were suitably affected, and a skin conductance response ensued. This is important because it shows, once again, that patients can generate skin conductance responses under certain conditions but not others. It is apparent that they respond to stimuli that are occurring now—a light, a sound, a loss, a gain—but that they will not respond if the trigger was a mental representation of something related to the stimulus but not available in direct perception. At first glance, one might describe their predicament by the saying "out of sight, out of mind," with which Patricia Goldman-Rakic aptly captures the working-memory defect resulting from dorsolateral frontal dysfunction. But we know that in these patients "out of sight" may be "still in mind," only it does not matter. Perhaps a better description for our patients is "out of sight and in mind, but never mind."

Within a number of card-turns into the game, something quite intriguing also began to happen in the normal subjects. In the period immediately preceding their selection of a card from a bad deck, that is, while the subjects were deliberating or had deliberated to pick from what the experimenter knew to be a bad deck, a skin conductance response was generated, and its magnitude increased as the game continued. In other words, the brains of the normal subjects were gradually learning to predict a bad outcome, and were signaling the relative badness of the particular deck before the actual card-turning.[4]

The fact that normal subjects did not show these responses when the game started, the fact that the responses were acquired from experience, over time, and the fact that their magnitude kept growing as more negative and positive experiences accrued were all strong indications that the brains of the normal subjects were learning something important about the situation and trying to signal, in anticipatory fashion, what would not be good for the future ahead.

If the presence of these responses in the normal subjects was fascinating, what we saw in the recordings of the frontally damaged patients was even more so: *the patients showed no anticipatory responses whatsoever*, no sign that their brains were developing a prediction for a negative future outcome.

Perhaps more than any other result, this one demonstrates both the predicament and a significant part of the underlying neuropathology in these patients. The neural systems that would have allowed them to learn what to avoid or prefer are malfunctioning, and are unable to develop responses suitable to a new situation.

We do not know yet how the prediction for negative future outcome develops in our gambling experiment. One wonders whether subjects make a cognitive estimate of badness versus goodness for each deck, and automatically connect that hunch with a somatic state signifying badness, which can, in turn, start operating as an alarm signal. In this formulation, reasoning, a cognitive estimate, precedes somatic signaling; but somatic signaling is still the critical component to implementation, because we know that patients cannot operate "normally" even if they know which decks are bad and which decks are good.

But there is one other possibility. It posits that a covert, nonconscious estimate precedes any cognitive process on the topic. The prefrontal networks would hone in on the ratio of badness versus goodness for each deck, on the basis of the frequency of bad and good somatic states experienced *after* punishment and reward. Helped by this automated sorting-out, the subject would be "helped into thinking" of the likely badness or goodness of each deck, that is, be guided into a theory about the game. Basic body regulatory systems would prepare the ground for conscious, cognitive processing. Without such preparation, the realization of what is good and what is bad would either never arrive, or would arrive too late and be too little.

Ten

The Body-Minded Brain

NO BODY, NEVER MIND

"HIS BODY HAS gone to his brain" is one of the least known among Dorothy Parker's celebrated epigrams. We can be certain that Miss Parker's unbridled wit was never concerned with neurobiology, that she was not referring to William James, and that she had not heard of George Lakoff or Mark Johnson, a linguist and a philosopher who have certainly had the body in their minds.[1] But her quip might provide some relief to readers impatient with my musings on the body-minded brain. In the pages ahead I return to the idea that the body provides a ground reference for the mind.

Imagine yourself walking home alone, around midnight, in whatever metropolis it is that you still walk home in, and realizing all of a sudden that somebody is persistently following you not far behind. In commonsense discourse, this is what happens: Your brain detects the threat; conjures up a few response options; selects one; acts on it; thus reduces or eliminates risk. As we have seen in the discussion on emotions, however, things are more complicated than that. The

neural and chemical aspects of the brain's response cause a profound change in the way tissues and whole organ systems operate.
The energy availability and the metabolic rate of the entire organism
are altered, as is the readiness of the immune system; the overall
biochemical profile of the organism fluctuates rapidly; the skeletal
muscles that allow the movement of head, trunk, and limbs contract;
and signals about all these changes are relayed back to the brain,
some via neural routes, some via chemical routes in the bloodstream,
so that the evolving state of the body proper, which has modified
continuously second after second, will affect the central nervous
system, neurally and chemically, at varied sites. The net result of
having the brain detect danger (or any similarly exciting situation) is
a profound departure from business as usual, both in restricted
sectors of the organism ("local" changes) and in the organism as a
whole ("global" changes). Most importantly, the changes occur in
both brain and body proper.

Despite the many examples of such complex cycles of interaction
now known, body and brain are usually conceptualized as separate,
in structure and function. The idea that it is the entire organism
rather than the body alone or the brain alone that interacts with the
environment often is discounted, if it is even considered. Yet when
we see, or hear, or touch or taste or smell, body proper *and* brain
participate in the interaction with the environment.

Think of viewing a favorite landscape. Far more than the retina
and the brain's visual cortices are involved. One might say that while
the cornea is passive, the lens and the iris not only let light through
but also adjust their size and shape in response to the scene before
them. The eyeball is positioned by several muscles, so as to track
objects effectively, and the head and neck move into optimal position. Unless these and other adjustments take place, you actually
may not see much. All of these adjustments depend on signals going
from brain to body and on related signals going from body to brain.

Subsequently, signals about the landscape are processed inside
the brain. Subcortical structures such as the superior colliculi are
activated; so are the early sensory cortices and the various stations of

the association cortex and the limbic system interconnected with them. As knowledge pertinent to the landscape is activated internally from dispositional representations in those various brain areas, the rest of the body participates in the process. Sooner or later, the viscera are made to react to the images you are seeing, and to the images your memory is generating internally, relative to what you see. Eventually, when a memory of the seen landscape is formed, that memory will be a neural record of many of the organismic changes just described, some of which happen in the brain itself (the image constructed for the outside world, together with the images constituted from memory) and some of which happen in the body proper.

Perceiving the environment, then, is not just a matter of having the brain receive direct signals from a given stimulus, let alone receiving direct pictures. The organism actively modifies itself so that the interfacing can take place as well as possible. The body proper is not passive. Perhaps no less important, the reason why most of the interactions with the environment ever take place is that the organism requires their occurrence in order to maintain homeostasis, the state of functional balance. The organism continuously *acts* on the environment (actions and exploration did come first), so that it can propitiate the interactions necessary for survival. But if it is to succeed in avoiding danger and be efficient in finding food, sex, and shelter, it must *sense* the environment (smell, taste, touch, hear, see), so that appropriate actions can be taken in response to what is sensed. Perceiving is as much about acting on the environment as it is about receiving signals from it.

The idea that mind derives from the entire organism as an ensemble may sound counterintuitive at first. Of late, the concept of mind has moved from the ethereal nowhere place it occupied in the seventeenth century to its current residence in or around the brain—a bit of a demotion, but still a dignified station. To suggest that the mind itself depends on brain-body interactions, in terms of evolutionary biology, ontogeny (individual development), and current operation may seem too much. But stay with me. What I am suggesting is that the mind arises from activity in neural circuits, to be sure, but many of those circuits were shaped in evolution by functional requisites of the organism, and that a normal mind will happen only if those circuits contain basic representations of the organism, and if they continue monitoring the states of the organism in action. In brief, neural circuits represent the organism continuously, as it is perturbed by stimuli from the physical and sociocultural environments, and as it acts on those environments. If the basic topic of those representations were not an organism anchored in the body, we might have some form of mind, but I doubt that it would be the mind we do have.

I am not saying that the mind is in the body. I am saying that the body contributes more than life support and modulatory effects to the brain. It contributes a *content* that is part and parcel of the workings of the normal mind.

Let us return to the example of your midnight walk home. Your brain has detected a threat, namely the person following you, and initiates several complicated chains of biochemical and neural reactions. Some of the lines in this internal screenplay are written in the body proper, and some are written in the brain itself. Yet you do not neatly differentiate between what goes on in your brain and what goes on in your body, even if you are an expert on the underlying neurophysiology and neuroendocrinology. You will be aware that you are in danger, that you are now quite alarmed and perhaps should walk faster, that you are walking faster, and that—one hopes—you are finally out of danger. The "you" in this episode is of one piece: in fact, it is a very real mental construction I will call "self" (for lack of a better word), and it is based on activities throughout your entire organism, that is, in the body proper and in the brain.

A sketch of what I think is necessary for the neural basis of self appears below, but I must immediately say that the self is a repeatedly reconstructed biological state; it is *not* a little person, the infamous homunculus, inside your brain contemplating what is going on. I mention that little man again only to let you know that I am not relying on him. It does not help to invoke a homunculus doing any seeing or thinking or whatever in your brain, because the natural question is whether the brain of that homunculus also has a little person in his brain doing his seeing and thinking, and so on ad infinitum. That particular explanation, which poses the problem of infinite regress,* is no explanation at all. I must point out also that having a self, a single self, is quite compatible with Dennett's notion that we have no Cartesian theater in some part of our brains. There is, to be sure, one self for each organism, except in those situations in which brain disease has created more than one (as happens in multiple personality disorder), or diminished or abolished the one normal self (as happens in certain forms of anosognosia and in certain types of seizure). But the self, that endows our experience with subjectivity, is not a central knower and inspector of everything that happens in our minds.

For the biological state of self to occur, numerous brain systems must be in full swing, as must numerous body-proper systems. If you were to cut *all* the nerves that bring brain signals to the body proper, your body state would change radically, and so consequently would your mind. Were you to cut *only* the signals from the body proper to the brain, your mind would change too. Even partial blocking of brain-body traffic, as happens in patients with spinal cord injury, causes changes in mind state.[2]

There is a philosophical thought experiment known as "brain in a vat," which consists of imagining a brain removed from its body, maintained alive in a nutrient bath, and stimulated via its now dangling nerves in precisely the same way it would be stimulated

---

*I would actually prefer to call the problem infinite regress *in space*, to emphasize the point that the real trouble rests with the creation of a nest of Russian dolls, one inside the other looking at yet another.

were it inside the skull.[3] Some people believe such a brain would have normal mental experiences. Now, leaving aside the suspension of disbelief required for imagining such a thing (and for imagining all *Gedanken* experiments), I believe that this brain would not have a normal mind. The absence of stimuli going *out* into the body-as-playing-field, capable of contributing to the renewal and modification of body states, would result in suspending the triggering and modulation of body states that, when represented back to the brain, constitute what I see as the bedrock of the sense of being alive. It might be argued that if it were possible to mimic, at the level of the dangling nerves, realistic configurations of inputs as if they were coming from the body, then the disembodied brain would have a normal mind. Well, that would be a nice and interesting experiment "to do" and I suspect the brain might indeed have *some* mind under those conditions. But what that more elaborate experiment would have done is create a body surrogate and thus confirm that "body-type inputs" are required for a normally minded brain after all. And what it would be unlikely to do is make the "body inputs" match in realistic fashion the variety of configurations which body states assume when those states are triggered by a brain engaged in making evaluations.

In brief, the representations your brain constructs to describe a situation, and the movements formulated as response to a situation, depend on mutual brain-body interactions. The brain constructs evolving representations of the body as it changes under chemical and neural influences. Some of those representations remain non-conscious, while others reach consciousness. At the same time, signals from the brain continue to flow to the body, some deliberately and some automatically, from brain quarters whose activities are never represented directly in consciousness. As a result, the body changes yet again, and the image you get of it changes accordingly.

While mental events are the result of activity in the brain's neurons, an early and indispensable story which brain neurons have to tell is the story of the body's schema and operation.

The primacy of the body as a theme applies to evolution: from simple to complex, for millions of years, brains have been first about the organism that owns them. To a lesser extent it applies also to the development of each of us as individuals so that at our beginning, there were first representations of the body proper, and only later were there representations related to the outside world; and to an even smaller but not negligible extent, to the *now*, as we construct the mind of the moment.

Making mind arise out of an organism rather than out of a disembodied brain is compatible with a number of assumptions.

First, when brains complex enough to generate not just motor responses (actions) but also mental responses (images in the mind) were selected in evolution, it was probably because those mental responses enhanced organism survival by one or all of the following means: a greater appreciation of external circumstances (for instance, perceiving more details about an object, locating it more accurately in space, and so on); a refinement of motor responses (hitting a target with greater precision); and a prediction of future consequences by way of imagining scenarios and planning actions conducive to achieving the best imagined scenarios.

Second, since *minded* survival was aimed at the survival of the whole organism, the primordial representations of the minding brain had to concern the body proper, in terms of its structure and functional states, including the external and internal actions with which the organism responded to the environment. It would not have been possible to regulate and protect the organism without representing its anatomy and physiology in both basic and *current* detail.

Developing a mind, which really means developing representations of which one can be made conscious as images, gave organisms a new way to adapt to circumstances of the environment that could not have been foreseen in the genome. The basis for that adaptability probably began by constructing images of the body proper in operation, namely images of the body as it responded to the environment externally (say, using a limb) and internally (regulating the state of viscera).

If ensuring survival of the body proper is what the brain first evolved for, then, when minded brains appeared, they began by minding the body. And to ensure body survival as effectively as possible, nature, I suggest, stumbled on a highly effective solution: *representing the outside world in terms of the modifications it causes in the body proper*, that is, representing the environment by modifying the primordial representations of the body proper whenever an interaction between organism and environment takes place.

What and where is this primordial representation? I believe it encompasses: (1) the representation of states of biochemical regulation in structures of the brain stem and hypothalamus; (2) the representation of the viscera, including not only the organs in the head, chest and abdomen, but also the muscular mass and the skin, which functions as an organ and constitutes the boundary of the organism, the supermembrane which encloses us as a unit; and (3) the representation of the musculoskeletal frame and its potential movement. These representations, which, as I indicated earlier, in chapters 4 and 7, are distributed over several brain regions, must be coordinated by neuron connections. I suspect that the representation of the skin and musculoskeletal frame may play an important role in securing that coordination, as explained below.

The first idea that comes to mind when we think of the skin is that of an extended sensory sheet, turned to the outside, ready to help us construct the shape, surface, texture, and temperature of external objects, through the sense of touch. But the skin is far more than that. First, it is a key player in homeostatic regulation: it is controlled by direct autonomic neural signals from the brain, and by chemical signals from numerous sources. When you blush or turn pale, the blushing or pallor happens in the "visceral" skin, not really in the skin you know as a touch sensor. In its visceral role—the skin is, in effect, the largest viscus in the entire body—the skin helps regulate body temperature by setting the caliber of the blood vessels housed in the thick of it, and helps regulate metabolism by mediating changes of ions (as when you perspire). The reason why people die from burns is not because they lose an integral part of their sense of touch. They die because the skin is an indispensable viscus.

My idea is that the brain's somatosensory complex, especially that of the right hemisphere in humans, represents our body structure by reference to a body schema where there are midline parts (trunk, head), appendicular parts (limbs), and a body boundary. A representation of the skin might be the natural means to signify the body boundary because it is an interface turned both to the organism's interior and to the environment with which the organism interacts.

This dynamic map of the overall organism anchored in body schema and body boundary would not be achieved in one brain area alone but rather in several areas by means of temporally coordinated patterns of neural activity. The indistinctly mapped representation of body operations at the level of brain stem and hypothalamus (where the topographic organization of neural activity is minimal) would be connected to brain regions where more and more topographic organization of signaling is available—the insular cortices, and the somatosensory cortices known as S1 and S2.[4] The sensory representation of all parts with a potential for movement would be connected to varied sites and levels of the motor system whose activity can cause muscular activity. In other words, the dynamic set of maps I have in mind is "somato-motor."

That the structures outlined above exist is not in question. I cannot guarantee, though, that they operate as I describe or that they play the role I suspect they play. But my hypothesis can be investigated. In the meantime, consider that if we did not have something like this device available, we would never be able to indicate the approximate location of pain or discomfort anywhere in our body, however imprecise we may be when we do; we would not be able to detect heaviness in the legs after standing for a long time, or queasiness in the abdomen, or the nausea and fatigue that signal jet lag and which we "localize" to just about the whole body.

Let us assume that my hypothesis might be supported, and discuss some of its implications. The first is that most interactions with the environment happen at *a place* within the body boundary, whether touch or another sense is being engaged, because sense organs exist at a location in the vast geographic map of this boundary. Signaling that involves an organism's interactions with its external surroundings may well be processed by reference to the overall map of the body boundary. A special sense, such as vision, is processed at *a special place* within the body boundary, in this case the eyes.

Signals from the outside are thus *double*. Something you see or hear excites the special sense of sight or sound as a "nonbody" signal, but it also excites a "body" signal hailing from the place in the skin where the special signal entered. As the special senses are engaged, they produce a dual set of signals. The first set comes from the body, originating in the particular location of the special sense organ (the eye in seeing, the ear in hearing), and is conveyed to the somatosensory and motor complex which dynamically represents the entire body as a functional map. The second set comes from the special organ itself and is represented in the sensory units appropriate to the sensory modality. (For seeing, these include the early visual cortices and the superior colliculi.)

This arrangement would have a practical consequence. When you see, you do not just see: *you feel you are seeing something with your eyes.* Your brain processes signals about your organism's being engaged at a specific place on the body reference map (such as the eyes and their controlling muscles), and about the visual specifics of whatever it is that excites your retinas.

I suspect that the knowledge that organisms acquired from touching an object, from seeing a landscape, from hearing a voice, or from moving in space along a given trajectory was represented by reference to the body in action. In the beginning, there was no touching, or seeing, or hearing, or moving along by itself. There was, rather, *a feeling of the body* as it touched, or saw, or heard, or moved.

To a considerable extent, this arrangement would have been maintained. It is appropriate to describe our visual perception as a "feeling of the body as we see," and we certainly "feel" we are seeing with our eyes rather than with our forehead. (We also "know" that we see with the eyes because if we close them, off go the visual images. But that inference is not equivalent to the natural feeling of seeing with the eyes.) It is true that the attention allocated to the visual processing itself does tend to make us partly unaware of the body. However, if pain, discomfort, or emotion set in, attention can be focused instantly on body representations, and the body feeling moves out of the background and into center stage.

We are actually far more aware of the overall state of the body than we usually admit, but it is apparent that as vision, hearing, and touch evolved, the attention usually allocated to their component of overall perception increased accordingly; thus the perception of the body proper more often than not was left precisely where it did, and does, the best job: *in the background*. This idea is consistent with the fact that in simple organisms, in addition to the forerunner of a body sense, which derives from the organisms' entire body boundary, or "skin," there are forerunners of the special senses (vision, hearing, touch), as can be gleaned from the way the *entire* body boundary may respond (to light, vibration, and mechanical contacts, respectively). Even in an organism without a visual system, one can encounter a forerunner of vision in the form of whole-body photosensitivity: The intriguing idea is that when photosensitivity is harnessed by a specialized part of the body (the eye), that very part itself has a specific *place* in the overall schema of the body. (The idea that eyes evolved from light-sensitive patches is Darwin's. Nicholas Humphrey has used the idea similarly.[5])

In most instances of regular perceptual operation, the somatosensory system and the motor system are engaged simultaneously along with the sensory system or systems appropriate to the objects being perceived. This is true even when the appropriate sensory system happens to be the exteroceptive, or externally oriented, component of the somatosensory system. When you touch an object, there are thus two sets of local signals from your skin. One is concerned with the object's shape and texture; the other is concerned with the places on the body being activated by contact with the object and by the arm and hand movement. Add to all this that since the object may generate a subsequent body reaction, relative to its emotional value, the somatosensory system is again engaged, shortly after that reaction. The near inevitability of body processing, regardless of what it is that we are doing or thinking, should be apparent. Mind is probably not conceivable without some sort of *embodiment,* a notion that figures prominently in the theoretical proposals of George Lakoff, Mark Johnson, Eleanor Rosch, Francisco Varela, and Gerald Edelman.[6]

I have discussed this idea with a diverse public and, if my experience is any indication, most readers will be comfortable with this account, but a few will find it extreme or wrong. I have listened carefully to the skeptics and learned that their main objection comes from what they sense as a lack of current, prevalent experience of anything bodily as they go about their own thinking. I do not see this as a problem, however, since I am not suggesting that body representations dominate the landscape of our mind (moments of emotional upheaval excepted). As far as the current moment is concerned, my idea is that images of body state are in the background, usually unattended but ready to spring forward. Moreover, the weight of my idea concerns the *history of development* of brain/mind processes rather than the current moment. I believe images of body state were indispensable, as building blocks and scaffolding, for what exists now. Without a doubt, however, what exists now is dominated by non-body images.

Another source of skepticism is the notion that the body was indeed relevant in the evolution of the brain but is so thoroughly and permanently "symbolized" in brain structure that it no longer needs to be "in the loop." Now this is certainly an extreme view. I agree that the body is well "symbolized" in brain structure, and that "symbols" of body may be used "as if" they were current body signals. But I prefer to think that the body remains "in the loop" for all the reasons I outlined. We simply have to wait for additional evidence to decide on the merits of the idea proposed here. In the meantime, I ask the skeptics to be patient.

THE BODY AS GROUND REFERENCE

Primordial representations of the body proper in action would offer a spatial and temporal framework, a metric on which other representations could be grounded. The representation of what we now construct as a space with three dimensions would be engendered in the brain, on the basis of the body's anatomy and patterns of movement in the environment.

While there is an external reality, what we know of it would come through the agency of the body proper in action, via representations of its perturbations. We would never know how faithful our knowledge is to "absolute" reality. What we need to have, and I believe we do have, is a remarkable consistency in the constructions of reality that our brains make and share.

Consider our idea of cats: we must construct *some* picture of how our organisms tend to be modified by a class of entities that we will come to know as cats, and we need to do that consistently, both individually and in the human collectives in which we live. Those systematic, consistent representations of cats are real in themselves. Our minds are real, our images of cats are real, our feelings about cats are real. It is just that such a mental, neural, biological reality happens to be *our* reality. Frogs or birds looking at cats see them differently, and so do cats themselves.

Perhaps most important, primordial representations of the body proper in action might play a role in consciousness. They would provide a core for the neural representation of self and thus provide a natural reference for what happens to the organism, inside or outside its boundary. The grounding reference in the body proper obviates the need to attribute to a homunculus the production of subjectivity. Instead there would be successive organism states, each neurally represented anew, in multiple concerted maps, moment by moment, and each anchoring the self that exists at any one moment.

THE NEURAL SELF

I am immensely interested in the subject of consciousness and am convinced that neurobiology can begin to approach the subject. Some philosophers (among them John Searle, Patricia Churchland, and Paul Churchland) have urged neurobiologists to study consciousness, and both philosophers and neurobiologists (Francis Crick, Daniel Dennett, Gerald Edelman, Rodolfo Llinás, among others) have begun to theorize about it.[7] But since this book is not about consciousness, I will confine my comments to one aspect that is pertinent to the discussion on images, feelings, and somatic markers. It concerns the neural basis of the self, the understanding of which might shed some light on the process of subjectivity, a key feature of consciousness.

I must first clarify what I mean by self, and to do so I offer an observation that I have made repeatedly in many patients struck by neurological disease. When a patient develops an inability to recognize familiar faces, or see color, or read, or when patients cease to recognize melodies, or understand speech, or produce speech, the description they offer of the phenomenon, with rare exceptions, is that something is happening to them, something new and unusual which they can observe, puzzle over, and often describe, in insightful and concrete ways. Curiously, the theory of mind implicit in those descriptions suggests that they "locate" the problem to a part of their persons which they are surveying from the vantage point of their selfhood. The frame of reference is not different from the one they would use were they referring to a problem with their knees or elbows. As I indicated, there are some rare exceptions; some patients with severe aphasia may not be as keenly aware of their defect and will not offer a clear account of the events in their minds. But usually, even the precise moment when the defect began is well remembered (these conditions often begin acutely). Countless times I have heard patients describe their experience of the dreaded moment when a brain lesion started and a cognitive or motor impairment set in: "My God, what is happening to *me*?" is a common utterance. None of these complicated defects is ever referred to a vague entity, or to the person next door. They are happening to the self.

Now let me tell you what occurs among patients with the complete form of anosognosia discussed earlier. Neither in my experience nor in any account I have read do they give an account comparable to those of the patients described in the preceding paragraph. Not one says, in effect, "God, how bizarre it is that I no longer feel any part of my body and that all that is left of me is my mind." Not one can tell you *when* the trouble started. They do not know, unless they are told. Unlike the patients to whom I referred above, none of the anosognosics can refer the trouble to the self.

Even more curious is the observation that patients with only a partial impairment of body sense can refer the problem to the self. This happens in patients with transient anosognosia or with what is known as asomatognosia. A telling example occurred in a patient who had a temporary loss of the sense of her entire body frame and body boundary (both left and right sides) but was nonetheless well aware of her visceral functions (breathing, heartbeat, digestion) and who could characterize her condition as a disquieting loss of part of her body but not of her "being." She still had a self—in fact, quite an alarmed self—whenever a new episode of partial loss of body sense occurred. The patient had seizures, which arose in a small but strategically located lesion in the right hemisphere, at the intersection of the several somatosensory maps I discussed previously; her lesion spared the anterior insula, the region that I believe holds the key to visceral sense; antiseizure medication promptly abolished the episodes.

My interpretation of the condition of complete anosognosics is that the damage they have sustained has partially demolished the substrate of the neural self. The state of self that they are able to construct is thus impoverished because of their impaired ability to process current body states. It relies on old information, which grows older by the minute.

. . .

The focus on self does not mean that I am talking about self-consciousness, since I see self and the subjectivity it begets as necessary for consciousness in general and not just for self-consciousness. Nor does an interest in the self mean that other features of consciousness are less important or less approachable by neurobiology. The process of making images, and the wakefulness and arousal which are necessary for the formation of those images, are just as relevant as the self, whom we experience as knower and owner of those images. Yet the problem of the neural basis for the self, and that of the neural basis for the formation of images, are not at the same level, cognitively or neurally. You cannot have a self without wakefulness, arousal, and the formation of images, but technically you can be awake and aroused and have images formed in sectors of your brain and mind, while having a compromised self. In extreme cases, the pathological alteration of wakefulness and arousal causes stupor, vegetative state, and coma, conditions in which the self vanishes entirely, as Fred Plum and Jerome Posner have shown in a classical description.[8] But there can be pathological alterations of the self without disruption of those basic processes, as patients with some type of seizure or complete anosognosia can demonstrate.

One other word of qualification before we proceed: In using the notion of self, I am in no way suggesting that *all* the contents of our minds are inspected by a single central knower and owner, and even less that such an entity would reside in a single brain place. I am saying, though, that our experiences tend to have a consistent perspective, as if there were indeed an owner and knower for most, though not all, contents. I imagine this perspective to be rooted in a relatively stable, endlessly repeated biological state. The source of the stability is the predominantly invariant structure and operation of the organism, and the slowly evolving elements of autobiographical data.

The neural basis for the self, as I see it, resides with the continuous reactivation of at least two sets of representations. One set concerns representations of key events in an individual's autobiography, on the basis of which a notion of identity can be reconstructed repeatedly, by partial activation in topographically organized sensory maps. The set of dispositional representations describing any of our autobiographies concerns a large number of categorized facts that define our person: what we do, whom and what we like, what types of objects we use, which places and actions we most usually frequent and perform. You might picture this set of representations as the sort of file J. Edgar Hoover was expert at preparing except that it is held in the association cortices of many brain sites rather than in filing cabinets. Furthermore, over and above such categorizations, there are unique facts from our past that are constantly activated as mapped representations: where we live and work, what our job is precisely, our own name and the names of close kin and friends, of city and country, and so forth. Finally we have, in recent dispositional memory, a collection of recent events, along with their approximate temporal continuity, and we also have a collection of plans, a number of imaginary events we intend to make happen, or expect to happen. The plans and imaginary events constitute what I call a "memory of the possible future." It is held in dispositional representations just like any other memory.

In brief, the endless reactivation of updated images about our identity (a combination of memories of the past and of the planned future) constitutes a sizable part of the state of self as I understand it.

The second set of representations underlying the neural self consists of the primordial representations of an individual's body, to which I alluded earlier: not only what the body has been like in general, but also what the body has been like *lately*, just before the processes leading to the perception of object X (this is an important point: as you will see below, I believe subjectivity depends in great part on the changes that take place in the body state during and after the processing of object X). Of necessity, this encompasses background body states and emotional states. The collective representation of the body constitute the basis for a "concept" of self, much as a collection of representations of shape, size, color, texture, and taste can constitute the basis for the concept of orange. Early body signals, in both evolution and development, helped form a "basic concept" of self; this basic concept provided the ground reference for whatever else happened to the organism, including the current body states that were incorporated *continuously* in the concept of self and promptly became past states. (They were the antecedent and foundation for the notion of self as formulated by Jerome Kagan.[9]) What is happening to us *now* is, in fact, happening to a concept of self based on the past, including the past that was current only a moment ago.

At each moment the state of self is constructed, from the ground up. It is an evanescent reference state, so continuously and consistently *re*constructed that the owner never knows it is being *re*made unless something goes wrong with the remaking. The background feeling now, or the feeling of an emotion now, along with the nonbody sensory signals now, happen to the concept of self as instantiated in the coordinated activity of multiple brain regions. But our self, or better even, our metaself, only "learns" about that "now" an instant later. Pascal's statements on past, present, and future, with which I opened chapter 8, capture this essence in lapidary fashion. Present continuously becomes past, and by the time we take stock of it we are in another present, consumed with planning the future, which we do on the stepping-stones of the past. The present is never here. We are hopelessly late for consciousness.

Finally, let me turn to perhaps the most critical issue in this discussion. By which legerdemain do an image of object X and a state of self, both of which exist as momentary activations of topographically organized representations, generate the subjectivity which characterizes our experiences? Let me preview the answer by saying that it depends on the brain's creation of a description, and on the imagetic display of that description. As images corresponding to a newly perceived entity (e.g., a face) are formed in early sensory cortices, the *brain reacts to those images*. This happens because signals arising in those images are relayed to several subcortical nuclei (e.g., the amygdala, the thalamus) and multiple cortical re-

gions; and because those nuclei and cortical regions contain dispositions for response to certain classes of signals. The end result is that
dispositional representations in nuclei and cortical regions are activated and, as a consequence, induce some collection of changes in
the state of the organism. In turn, those changes alter the body image
momentarily, and thus perturb the *current* instantiation of the concept of self.

Although the responding process implies knowledge, it certainly
does not imply that any brain component "knows" that responses are
being generated to the presence of an entity. When the organism's
brain generates a set of responses to an entity, the existence of a
representation of self does not make that self *know* that its corresponding organism is responding. The self, as described above,
cannot *know*. However, a process we could call "metaself" might
know, provided (1) the brain would create some kind of *description of
the perturbation of the state of the organism* that resulted from the
brain's responses to the presence of an image; (2) the description
would *generate an image of the process of perturbation,* and (3) the
image of the *self perturbed* would be displayed together or in rapid
interpolation with the image that triggered the perturbation. In
short, the description I am talking about concerns the *perturbation
of the organism's state,* as a result of the brain's responses to the image
of object X. The description does not use language although it can be
translated into language.

Having an image alone is not enough, even if we invoke attention
and awareness, because both attention and awareness are properties
of a self as it experiences images, i.e., as it is made aware of the
images to which it attends. Having both images *and* a self is not
sufficient either. To say that the image of an object is referred to the
images which constitute the self, or correlated with them, are not
particularly helpful statements. One would not understand what the
reference or correlation consist of, or what they achieve. How subjectivity would emerge from such a process would be entirely
mysterious.

Now consider the following possibilities. Consider, first of all, that

the brain possesses a third set of neural structures which is neither
the one which supports the image of an object nor the one that
supports the images of the self but is reciprocally interconnected
with both. In other words, the kind of third-party neuron ensemble,
which we have called a convergence zone, and which we have
invoked as the neural substrate for building dispositional representations all over the brain, in cortical regions as well as subcortical
nuclei.

Next, imagine that such a third-party ensemble receives signals
from both the representation of the object and the representations of
the self, *as the organism is perturbed by the representation of the
object.* In other words, imagine that the third-party ensemble is
building a *dispositional representation of the self in the process of
changing as the organism responds to an object.* There would be
nothing mysterious about this dispositional representation which
would be of precisely the same kind that the brain seems to be
exceedingly good at holding, making, and remodeling. Also, we know
that the brain has all the requisite information to build such a
dispositional representation: Shortly after we see an object and hold
a representation of it in early visual cortices, we also hold many
representations of the organism reacting to the object in varied
somatosensory regions.

The dispositional representation I have in mind is neither created
nor perceived by a homunculus, and, as is the case with all dispositions, it has the potential to reactivate, in the early sensory cortices to
which it is connected, an image of what the disposition is about: a
somatosensory image of the organism responding to a particular
object.

Finally consider that all ingredients I have described above—an
object that is being represented, an organism responding to the object
of representation, and a state of the self in the process of changing
because of the organism's response to the object—are held simultaneously in working memory and attended, side-by-side or in rapid
interpolation, in early sensory cortices. I propose that subjectivity
emerges during the latter step when the brain is producing not just

images of an object, not just images of organism responses to the object, but a third kind of image, that of an organism in the act of perceiving and responding to an object. I believe the subjective perspective arises out of the content of the third kind of image.

The minimal neural device capable of producing subjectivity thus requires early sensory cortices (including the somatosensory), sensory and motor cortical association regions, and subcortical nuclei (especially thalamus and basal ganglia) with convergence properties capable of acting as third-party ensembles.

This basic neural device does not require language. The metaself construction I envision is purely nonverbal, a schematic view of the main protagonists from a perspective external to both. In effect, the third-party view constitutes, moment-by-moment, a nonverbal narrative document of what is happening to those protagonists. The narrative can be accomplished without language, using the elementary representational tools of the sensory and motor systems in space and time. I see no reason why animals without language would not make such narratives.

Humans have available second order narrative capacities, provided by language, which can engender verbal narratives out of nonverbal ones. The refined form of subjectivity that is ours would emerge from the latter process. Language may not be the source of the self, but it certainly is the source of the "I."

I am not aware of another specific proposal for a neural basis of subjectivity, but since subjectivity is a key feature of consciousness it is appropriate to note, however briefly, where my proposal relates to others in this general area.

Francis Crick's hypothesis on consciousness is focused on the problem of image making and leaves out subjectivity altogether. Crick has not overlooked the problem of subjectivity. Rather, he has decided not to consider it at this time since he doubts it can be approached experimentally. His preference and caution are quite legitimate, but I worry that by postponing the consideration of subjectivity, we may be unable to interpret correctly the empirical data concerning the making and perception of images.

Daniel Dennett's hypothesis, on the other hand, pertains to the high end of consciousness, to the final products of the mind. He agrees that there is a self, but he does not address its neural basis and focuses instead on the mechanisms by which our experience of a stream-of-consciousness might be created. Interestingly, at that level of the process, he utilizes a notion of sequence construction (his Joycean virtual machine) that is not unlike the notion of image construction I utilize at a lower and earlier level. I am fairly certain, however, that my device for generating subjectivity is not Dennett's virtual machine.

My proposal shares an important characteristic with Gerald Edelman's views on the neural basis of consciousness, namely the acknowledgement of a biological self imbued with value. (Edelman has been virtually alone, among contemporary theorists, in the importance he has accorded to innate value in biological systems.) Edelman, however, restricts the biological self to subcortical homeostatic systems (whereas I incorporate it in factual, cortically based systems, and allow the products of their activity to become feelings). The processes I envision and the structures I propose to carry them are therefore different. Moreover, I am not certain of the degree of correspondence between my notion of subjectivity and Edelman's notion of primary consciousness.

William James, who thought that no reasonable psychology could question the existence of "personal selves," and who believed that the worst a psychology might do is rob those selves of significance, might be pleased to discover that today there are plausible if not yet proven hypotheses for the neural basis of the self.

Eleven

A Passion for
Reasoning

A T THE BEGINNING of this book I suggested that feelings are a
powerful influence on reason, that the brain systems required
by the former are enmeshed in those needed by the latter, and that
such specific systems are interwoven with those which regulate
the body.

The facts I have presented generally support these hypotheses,
but these are hypotheses nonetheless, offered in the hope that they
may attract further investigation and be subject to revision as new
findings appear. Feelings do seem to depend on a dedicated multicomponent system that is indissociable from biological regulation.
Reason does seem to depend on specific brain systems, some of
which happen to process feelings. Thus there may be a connecting
trail, in anatomical and functional terms, from reason to feelings to
body. It is as if we are possessed by a passion for reason, a drive that
originates in the brain core, permeates other levels of the nervous
system, and emerges as either feelings or nonconscious biases to
guide decision making. Reason, from the practical to the theoretical,
is probably constructed on this inherent drive by a process which
resembles the mastering of a skill or craft. Remove the drive, and you

will not acquire the mastery. But having the drive does not automatically make you a master.

Should these hypotheses be supported, are there sociocultural implications to the notion that reason is nowhere pure? I believe that there are, and that they are by and large positive.

Knowing about the relevance of feelings in the processes of reason does *not* suggest that reason is less important than feelings, that it should take a backseat to them or that it should be less cultivated. On the contrary, taking stock of the pervasive role of feelings may give us a chance of enhancing their positive effects and reducing their potential harm. Specifically, without diminishing the orienting value of normal feelings, one would want to protect reason from the weakness that abnormal feelings or the manipulation of normal feelings can introduce in the process of planning and deciding.

I do not believe that knowledge about feelings should make us less inclined to empirical verification. I only see that greater knowledge about the physiology of emotion and feeling should make us more aware of the pitfalls of scientific observation. The formulation I presented should not diminish our determination to control external circumstances to the advantage of individuals and society, or our resolve to develop, invent, or perfect the cultural instruments with which we can make the world better: ethics, law, art, science, technology. In other words, nothing in my formulation urges acceptance of things as they are. I must emphasize this point, since the mention of feelings often conjures up an image of self-oriented concern, of disregard for the world around, and of tolerance for relaxed standards of intellectual performance. That is, in effect, the very opposite of my view, and one less worry for those who, like the molecular biologist Gunther Stent, have been concerned, justly, that the overvaluing of feelings might result in less determination to uphold the Faustian contract that has brought progress to humanity.[1]

What worries me is the acceptance of the importance of feelings without any effort to understand their complex biological and sociocultural machinery. The best example of this attitude can be

found in the attempt to explain bruised feelings or irrational behavior by appealing to surface social causes or the action of neurotransmitters, two explanations that pervade the social discourse as
presented in the visual and printed media; and in the attempt to
correct personal and social problems with medical and nonmedical
drugs. It is precisely this lack of understanding of the nature of
feelings and reason (one of the hallmarks of the "culture of complaint"²) that is cause for alarm.

The idea of the human organism outlined in this book, and the
relation between feelings and reason that emerges from the findings
discussed here, do suggest, however, that the strengthening of rationality probably requires that greater consideration be given to the
vulnerability of the world within.

On a practical note, the role outlined for feelings in the making of
rationality has implications for some issues currently facing our
society, education and violence among them. This is not the place to
do justice to these issues but let me comment that educational
systems might benefit from emphasizing unequivocal connections
between current feelings and predicted future outcomes, and that
children's overexposure to violence, in real life, newscasts, or
through audiovisual fiction, downgrades the value of emotions and
feelings in the acquisition and deployment of adaptive social behavior. The fact that so much vicarious violence is presented without a
moral framework only compounds its desensitizing action.

DESCARTES' ERROR

It would not have been possible to present my side of this conversation without invoking Descartes as an emblem for a collection of
ideas on body, brain, and mind that in one way or another remain
influential in Western sciences and humanities. My concern, as you
have seen, is for both the dualist notion with which Descartes split
the mind from brain and body (in its extreme version, it holds less
sway) and for the modern variants of this notion: the idea, for
instance, that mind and brain are related, but only in the sense that

the mind is the software program run in a piece of computer hardware called brain; or that brain and body are related, but only in the sense that the former cannot survive without the life support of the latter.

What, then, was Descartes' error? Or better still, *which* error of Descartes' do I mean to single out, unkindly and ungratefully? One might begin with a complaint, and reproach him for having persuaded biologists to adopt, to this day, clockwork mechanics as a model for life processes. But perhaps that would not be quite fair and so one might continue with "I think therefore I am." The statement, perhaps the most famous in the history of philosophy, appears first in the fourth section of the *Discourse on the Method* (1637), in French (*"Je pense donc je suis"*); and then in the first part of the *Principles of Philosophy* (1644), in Latin (*"Cogito ergo sum"*).[3] Taken literally, the statement illustrates precisely the opposite of what I believe to be true about the origins of mind and about the relation between mind and body. It suggests that thinking, and awareness of thinking, are the real substrates of being. And since we know that Descartes imagined thinking as an activity quite separate from the body, it does celebrate the separation of mind, the "thinking thing" (*res cogitans*), from the nonthinking body, that which has extension and mechanical parts (*res extensa*).

Yet long before the dawn of humanity, beings were beings. At some point in evolution, an elementary consciousness began. With that elementary consciousness came a simple mind; with greater complexity of mind came the possibility of thinking and, even later, of using language to communicate and organize thinking better. For us then, in the beginning it was being, and only later was it thinking. And for us now, as we come into the world and develop, we still begin with being, and only later do we think. We are, and then we think, and we think only inasmuch as we are, since thinking is indeed caused by the structures and operations of being.

When we put Descartes' statement back where it belongs, we might wonder for a moment whether it might mean something different from what it has come to stand for. Might one read it instead as an acknowledgment of the superiority of conscious feeling and reasoning, without any firm commitment as to their origin, substance, or permanence? Might the statement also have served the clever purpose of accommodating religious pressures of which Descartes was keenly aware? The latter is a possibility, but there is no way of finding out for sure. (The inscription Descartes chose for his tombstone was a quote that he apparently used frequently: *"Bene qui latuit, bene vixit,"* from Ovid's *Tristia* 3.4.25. Translation: "He who hid well, lived well." A cryptic disclaimer of dualism, perhaps?) As for the former, on balance, I suspect Descartes *also* meant precisely what he wrote. As the famous words first appear, Descartes is rejoicing with the discovery of a proposition so undeniably true that no amount of skepticism will shake it:

> . . . and remarking that this truth *"I think, therefore I am"* was so certain and so assured that all the most extravagant suppositions brought forward by the sceptics were incapable of shaking it, I came to the conclusion that I would receive it without scruple as the first principle of the Philosophy for which I was seeking.[4]

Here Descartes was after a logical foundation for his philosophy, and the statement was not unlike Augustine's *"Fallor ergo sum"* (I am deceived therefore I am).[5] But just a few lines below, Descartes clarifies the statement unequivocally:

> From that I knew that I was a substance, the whole essence or nature of which is to think, and that for its existence there is no need of any place, nor does it depend on any material thing; so that this "me," that is to say, the soul by which I am what I am, is entirely distinct from body, and is even more easy to know than is the latter; and even if body were not, the soul would not cease to be what it is.[6]

This is Descartes' error: the abyssal separation between body and mind, between the sizable, dimensioned, mechanically operated, infinitely divisible body stuff, on the one hand, and the unsizable, undimensioned, un-pushpullable, nondivisible mind stuff; the suggestion that reasoning, and moral judgment, and the suffering that comes from physical pain or emotional upheaval might exist separately from the body. Specifically: the separation of the most refined operations of mind from the structure and operation of a biological organism.

Now, some may ask, why quibble with Descartes rather than with Plato, whose views on body and mind were far more exasperating, as can be discovered in the *Phaedo*? Why bother with this particular error of Descartes'? After all, some of his other errors sound more spectacularly wrong than this one. He believed that heat made the blood circulate, and that tiny, ever so fine particles of the blood distilled themselves into "animal spirits," which could then move muscles. Why not take him to task for either of those notions? The reason is simple: We have known for a long time that he was wrong on those particular points, and the questions of how and why the blood circulates have been answered to our complete satisfaction. That is not the case when we consider questions of mind, brain, and body, concerning which Descartes' error remains influential. For many, Descartes' views are regarded as self-evident and in no need of reexamination.

The Cartesian idea of a disembodied mind may well have been the source, by the middle of the twentieth century, for the metaphor of mind as software program. In fact, if mind can be separated from body, perhaps one can try to understand it without any appeal to neurobiology, without any need to be influenced by knowledge of neuroanatomy, neurophysiology, and neurochemistry. Interestingly and paradoxically, many cognitive scientists who believe they can investigate the mind without recourse to neurobiology would not consider themselves dualists.

There may be some Cartesian disembodiment also behind the thinking of neuroscientists who insist that the mind can be fully explained solely in terms of brain events, leaving by the wayside the rest of the organism and the surrounding physical and social environment—and also leaving out the fact that part of the environment is itself a product of the organism's preceding actions. I resist the restriction, not because the mind is not directly related to brain activity, since it obviously is, but rather because the restrictive formulation is unnecessarily incomplete; and humanly unsatisfactory. To say that mind comes from brain is indisputable, but I prefer to qualify the statement and consider the reasons why the brain's neurons behave in such a thoughtful manner. For the latter is, so far as I can see, the critical issue.

The idea of a disembodied mind also seems to have shaped the peculiar way in which Western medicine approaches the study and treatment of diseases (see the postscriptum). The Cartesian split pervades both research and practice. As a result, the psychological consequences of diseases of the body proper, the so-called real diseases, are usually disregarded and only considered on second thought. Even more neglected are the reverse, the body-proper effects of psychological conflict. How intriguing to think that Descartes did contribute to modifying the course of medicine, did help it veer from the organismic, mind-in-the-body approach, which prevailed from Hippocrates to the Renaissance. How annoyed Aristotle would have been with Descartes, had he known.

Versions of Descartes' error obscure the roots of the human mind in a biologically complex but fragile, finite, and unique organism; they obscure the tragedy implicit in the knowledge of that fragility, finiteness, and uniqueness. And where humans fail to see the inherent tragedy of conscious existence, they feel far less called upon to do something about minimizing it, and may have less respect for the value of life.

The facts I have presented about feelings and reason, along with others I have discussed about the interconnection between brain and body proper, support the most general idea with which I introduced the book: that the comprehensive understanding of the human mind requires an organismic perspective; that not only must the mind move from a nonphysical cogitum to the realm of biological tissue, but it must also be related to a whole organism possessed of integrated body proper and brain and fully interactive with a physical and social environment.

The truly embodied mind I envision, however, does not relinquish its most refined levels of operation, those constituting its soul and spirit. From my perspective, it is just that soul and spirit, with all their dignity and human scale, are now complex and unique states of an organism. Perhaps the most indispensable thing we can do as human beings, every day of our lives, is remind ourselves and others of our complexity, fragility, finiteness, and uniqueness. And this is of course the difficult job, is it not: to move the spirit from its nowhere pedestal to a somewhere place, while preserving its dignity and importance; to recognize its humble origin and vulnerability, yet still call upon its guidance. A difficult and indispensable job indeed, but one without which we will be far better off leaving Descartes' Error uncorrected.

Postscriptum

THE HUMAN HEART IN CONFLICT

" $\top$ HE POET'S VOICE need not merely be the record of man, it can be one of the props, the pillars to help him endure and prevail."[1] William Faulkner wrote these words about 1950, but they are just as applicable today. The audience he had in mind was that of his fellow writers, but he might as well have been exhorting those of us who study the brain and the mind: The scientist's voice need not be the mere record of life as it is; scientific knowledge can be a pillar to help humans endure and prevail. This book was written with the conviction that knowledge in general and neurobiological knowledge in particular have a role to play in human destiny; that if only we want it, deeper knowledge of brain and mind will help achieve the happiness whose yearning was the springboard for progress, two centuries ago, and will maintain the glorious freedom that Paul Éluard described in his poem "Liberté."[2]

In the same text cited above, Faulkner tells his fellow writers that they have "forgotten the problems of the human heart in conflict with itself, which alone can make good writing because only that is worth writing about, both the agony and the sweat." He asks them to leave no room in their workshops "for anything but the old verities and truths of the heart, the old universal truths lacking which any story is ephemeral and doomed—love and honor and pity and pride and compassion and sacrifice."

It is tempting and encouraging to believe, perhaps beyond Faulkner's meaning, that neurobiology not only can assist us with the comprehension and compassion of the human condition, but that in so doing it can help us understand social conflict and contribute to its alleviation. This is not to suggest that neurobiology can save the world, but simply that the gradual accrual of knowledge about human beings can help us find better ways for the management of human affairs.

For quite some time now, humans have been in a new, thoughtful phase of evolution, in which their minds and brains can be both servants and masters of their bodies and of the societies they constitute. Of course, there are risks when brains and minds that came from nature decide to play sorcerer's apprentice and influence nature itself. But there are also risks in not taking the challenge and not attempting to minimize suffering. There are, in fact, enormous risks in not doing anything. Doing just what comes naturally can only please those who are unable to imagine better worlds and better ways, those who believe they are already in the best of all possible worlds.[3]

MODERN NEUROBIOLOGY AND THE IDEA OF MEDICINE

There is something paradoxical about the conceptualization of medicine and about its practitioners in our culture. A number of physicians have interests in the humanities, from the arts to literature to philosophy. Some surprising number of them have become poets, novelists, and playwrights, of eminence, and several have reflected with depth on the human condition and dealt perceptively with its psychological, social, and political dimensions. And yet the medical schools they have come from largely ignore those human dimensions as they concentrate on the physiology and pathology of the body proper. Western medicine, especially medicine in the United States, came to glory through the expansion of internal medicine and surgical subspecialties, both of which had as targets the diagnosis and treatment of diseased organs and systems throughout the body. The brain (more precisely, the central and peripheral nervous systems) was included in the effort since it was one such organ system. But its most precious product, the mind, was of little concern to mainstream medicine and, in fact, has not been the principal focus of the specialty that emerged from the study of brain diseases: neurology. It is perhaps no accident that American neurology began as a subspecialty of internal medicine and gained independence only in the twentieth century.

The net result of this tradition has been a remarkable neglect of the mind as a function of the organism. Few medical schools, to this day, offer their students any formal instruction on the normal mind, instruction that can come only from a curriculum strong in general psychology, neuropsychology, and neuroscience. Medical schools do offer studies of the sick mind encountered in mental diseases, but it is indeed astonishing to realize that students learn about psychopathology without ever being taught normal psychology.

There are several reasons behind this state of affairs, and I submit that most of them derive from a Cartesian view of humanity. For the past three centuries, the aim of biological studies and of medicine has been the understanding of the physiology and pathology of the body proper. The mind was out, largely left as a concern for religion and philosophy, and even after it became the focus of a specific discipline, psychology, it did not begin to gain entry into biology and medicine until recently. I am aware of commendable exceptions to this panorama, but they simply reinforce the idea I am giving of the general situation.

The result of all this has been an amputation of the concept of humanity with which medicine does its job. It should not be surprising that, by and large, the consequences of diseases of the body proper on the mind are a second thought, or no thought at all. Medicine has been slow to realize that how people feel about their medical condition is a major factor in the outcome of treatment. We still know very little about the placebo effect, through which patients respond beneficially in excess of what a given medical intervention would lead one to expect. (The placebo effect can be assessed by investigating the effect of tablets or injections which, unbeknownst to the patient, contain no active pharmacological ingredient and are thus presumed to have no influence whatever, positive or negative.) For instance, we do not know who is more likely to respond with a placebo effect, or if all of us can. We also do not know how far the placebo effect can go and how close to the effect of the real thing it can get. We know little about how to enhance the placebo effect. And we have no idea about the degree of error the placebo effect has created for so-called double-blind studies.

The fact that psychological disturbances, mild or strong, can cause diseases of the body proper is finally beginning to be accepted, but the circumstances in which they can, and the degree to which they can, remain unstudied. Of course our grandmothers knew all about this: they could tell us how grief, obsessive worry, excessive anger, and so forth would damage hearts, give ulcers, destroy complexions, and make one more prone to infections. But that was all too "folksy," too "soft" as far as science was concerned, and so it was. It took a long time for medicine to begin discovering that the basis for such human wisdom was worth considering and investigating.

The Cartesian-based neglect of the mind in Western biology and medicine has had two major negative consequences. The first is in the realm of science. The effort to understand the mind in general biological terms has been retarded by several decades, and it is fair to say that it has barely begun. Better late than never, that is for sure, but the delay means also that the potential impact that a deep understanding of the biology of mind might have had in human affairs has so far been lost.

The second negative consequence has to do with the effective diagnosis and treatment of human disease. It is of course true that all great physicians have been those men and women who are not only well versed in the hard-core physiopathology of their time, but are equally at ease, mostly through their own insight and accumulated wisdom, with the human heart in conflict. They have been expert diagnosticians and miracle workers, because of a *combination* of knowledge and talent. Yet we would be deluding ourselves if we thought that the standard of medical practice in the Western world is that of those notable physicians we all have known. A distorted view of the human organism, combined with the overwhelming growth of knowledge and the need for subspecialization, conspires to increase the inadequacy of medicine rather than reduce it. Medicine hardly needed the additional problems that have come from its economics, but it is getting those too, and they are certain to worsen medical performance.

The problem with the rift between body and mind in Western medicine has not yet been articulated by the public at large, although it seems to have been detected. I even suspect that the success of some "alternative" forms of medicine, especially those rooted in non-Western traditions of medicine, is probably a compensatory response to the problem. There is something to be admired and learned in those alternative forms of medicine, but unfortunately, regardless of how humanly adequate they may be, what they offer is not enough to deal effectively with human disease. In all fairness, we have to recognize that even mediocre Western medicine does solve a remarkable number of problems, quite decisively. But alternative forms of medicine do point to a blatant area of weakness in Western medical tradition that should be corrected scientifically, within scientific medicine itself. If, as I believe, the current success of alternative medicine is a symptom of public dissatisfaction with traditional medicine's inability to consider humans as a whole, then this dissatisfaction is likely to grow in the years ahead, as the spiritual crisis of Western society deepens.

The proclamation of bruised feelings, the desperate plea for the correction of individual pain and suffering, the inchoate cry for the loss of a never-achieved sense of inner balance and happiness to which most humans aspire are not likely to diminish soon.[4] It would be foolish to ask medicine alone to heal a sick culture, but it is just as foolish to ignore that aspect of human disease.

A NOTE ON THE LIMITS OF NEUROBIOLOGY NOW

Throughout this book I have spoken about accepted facts, disputed facts, and interpretations of facts; about ideas shared or not shared by many of us in the brain-mind sciences; about things that are as I say, and things that may be as I say. The reader may have been surprised at my insistence that so many "facts" are uncertain and that so much of what can be said about the brain is best stated as working hypotheses. Naturally, I wish I could say that we know with certainty how the brain goes about the business of making mind, but I cannot—and, I am afraid, no one can.

I hasten to add that the lack of definitive answers on brain/mind matters is not a cause for despair, however, and is not to be seen as a sign of failure of the scientific fields now engaged in the effort. On the contrary, the spirit of the troops is high since the rate at which new findings are accruing is greater than ever. The lack of precise and comprehensive explanations does not indicate an impasse. There is reason to believe that we will arrive at satisfactory explanations, although it would be foolhardy to set a date for the arrival, and even more so to say that they are around the corner. If there is any cause for worry, it comes not from a lack of progress but rather from the torrent of new facts that neuroscience is delivering and the threat that they may engulf the ability to think clearly.

If we have this wealth of new facts, you may ask, why are definitive answers not available? Why can we not give a precise and comprehensive account of how we see and, more important, how there is a self doing that seeing?

The principal reason for the delay—one might even say the only reason—is the sheer complexity of the problems for which we need answers. It is obvious that what we want to understand depends largely on the operation of neurons, and we do have a substantial knowledge about the structure and function of those neurons, all the way down to the molecules constituting them and making them do what they do best: fire, or engage in patterns of excitation. We even know something about the genes that make those neurons be and operate in a certain fashion. But clearly, human minds depend on the overall firing of those neurons, as they constitute complicated assemblies ranging from local, microscopic scale circuits to macroscopic systems spanning several centimeters. There are several billion neurons in the circuits of one human brain. The number of synapses formed among those neurons is at least 10 trillion, and the length of the axon cables forming neuron circuits totals something on the order of several hundred thousand miles. (I thank Charles Stevens, a neurobiologist at the Salk Institute, for the informal estimate.) The product of activity in these circuits is a pattern of firing that is transmitted to another circuit. This circuit may or may not fire, depending on a host of influences, some local, provided by other neurons terminating in the vicinity, and some global, brought by chemical compounds such as hormones, arriving in the blood. The time scale for the firing is extremely small, on the order of tens of milliseconds—which means that within one second in the life of our minds, the brain produces millions of firing patterns over a large variety of circuits distributed over various brain regions.

It should be clear, then, that the secrets of the neural basis of mind cannot be discovered by unraveling all the mysteries of one single neuron, regardless of how typical that neuron might be; or by unraveling all the intricate patterns of local activity in a typical neuron circuit. To a first approximation, the elementary secrets of mind reside with the interaction of firing patterns generated by many neuron circuits, locally and globally, moment by moment, within the brain of a living organism.

There is not one simple answer to the brain/mind puzzle, but rather many answers, keyed to the myriad components of the nervous system at its many levels of structure. The approach to understanding those levels calls for various techniques and proceeds at various paces. Some of the work can be based on experiments in animals and tends to develop relatively fast. But other work can be carried out only in humans, with the appropriate ethical cautions and limitations, and the pace must be slower.

Some have asked why neuroscience has not yet achieved results as spectacular as those seen in molecular biology over the past four decades. Some have even asked what is the neuroscientific equivalent of the discovery of DNA structure, and whether or not a corresponding neuroscientific fact has been established. There is no such single correspondence, although some facts, at several levels of the nervous system, might be construed as comparable in practical value to knowing the structure of DNA—for instance, understanding what an action potential is all about. But the equivalent, at the level of mind-producing brain, has to be a *large-scale outline of circuit and system designs,* involving descriptions *at both microstructural and macrostructural levels.*

Should the reader find that the above justifications for the limits of our current knowledge seem insufficient, let me note two more. First, as I previously indicated, only a part of the circuitry in our brains is specified by genes. The human genome specifies the construction of our bodies in great detail, and that includes the overall design of the brain. But not all of the circuits actively develop and work as set by genes. Much of each brain's circuitry, at any given moment of adult life, is individual and unique, truly reflective of that particular organism's history and circumstances. Naturally, that does not make the unraveling of neural mysteries any easier. Second, each human organism operates in collectives of like beings; the mind and the behavior of individuals belonging to such collectives and operating in specific cultural and physical environments are not shaped merely by the activity-driven circuitries mentioned above, and even less are they shaped by genes alone. To understand in a satisfactory manner the brain that fabricates human mind and human behavior, it is necessary to take into account its social and cultural context. And that makes the endeavor truly daunting.

LEVERAGE FOR SURVIVAL

In some species, nonhuman and even nonprimate, in which memory, reasoning, and creativity are limited, there are, nonetheless, manifestations of complex social behavior whose neural control must be innate. Insects—ants and bees in particular—offer dramatic examples of social cooperation that might easily put to shame the United Nations General Assembly, most any day. Closer to home, mammals abound in such manifestations, and the behaviors of wolves, dolphins, and vampire bats, among other species, even suggest an ethical structure. It is apparent that humans possess some of those same innate mechanisms, and that such mechanisms are the likely basis for some ethical structures used by humans. The most elaborate social conventions and ethical structures by which we live, however, must have arisen culturally and been transmitted likewise.

If that is the case, one may wonder, what was the trigger for the cultural development of such strategies? It is likely that they evolved as a means to cope with the suffering experienced by individuals whose capacity to remember the past and anticipate the future had attained a remarkable development. In other words, the strategies evolved in individuals able to realize that their survival was threatened or that the quality of their post-survival life could be bettered. Such strategies could have evolved only in the few species whose brains were structured to permit the following: First, a large capacity to memorize categories of objects and events, and to memorize unique objects and events, that is, to establish dispositional representations of entities and events at the level of categories and at unique level. Second, a large capacity for manipulating the components of those memorized representations and fashioning new creations by means of novel combinations. The most immediately useful variety of those creations consisted of imagined scenarios, the anticipation of outcomes of actions, the formulation of future plans, and the design of new goals that can enhance survival. Third, a large capacity to memorize the new creations described above, that is, the anticipated outcomes, the new plans, and the new goals. I call those memorized creations "memories of the future."[5]

If enhanced knowledge of the experienced past and the anticipated future was the reason why social strategies had to be created to cope with suffering, we still must explain why suffering arose in the first place. And for that we must consider the biologically prescribed sense of pain as well as its opposite, pleasure. The curious thing is, of course, that the biological mechanisms behind what we now call pain and pleasure were also an important reason why the innate instruments of survival were selected and combined the way they were, in evolution, when there was no individual suffering or reason. This may simply mean that the same simple device, applied to systems with very different orders of complexity and in different circumstances, leads to different but related results. The immune system, the hypothalamus, the ventromedial frontal cortices, and the Bill of Rights have the same root cause.

Pain and pleasure are the levers the organism requires for instinctual and acquired strategies to operate efficiently. In all probability they were also the levers that controlled the development of social decision-making strategies. When many individuals, in social groups, experienced the painful consequences of psychological, social, and natural phenomena, it was possible to develop intellectual and cultural strategies for coping with the experience of pain and perhaps reducing it.

Pain and pleasure occur when we become conscious of body-state profiles that clearly deviate from the base range. The configuration of stimuli and of brain-activity patterns perceived as pain or pleasure are set a priori in the brain structure. They occur because circuits fire in a particular way, and those circuits exist because they were instructed genetically to form themselves in a particular way. Although our reactions to pain and pleasure can be modified by education, they are a prime example of mental phenomena that depend on the activation of innate dispositions.

We should distinguish at least two components in pain and pleasure. In the first, the brain plots the representation of a local body-
state change, which is referred to a part of the body. This is a
somatosensory perception in the proper sense. It derives from the
skin, or from a mucosa, or from part of an organ. The second
component of pain and pleasure results from a more general change
in body state, in fact an emotion. What we call pain or pleasure, for
example, is the name for a concept of a particular body landscape
that our brains are perceiving. The perception of that landscape is
modulated further in the brain by neurotransmitters and neuromodulators, which affect signal transmission and the operation of
the brain sectors concerned with representing the body. The release
of endorphins (the organism's own morphine), which bind to opioid
receptors (which are similar to those on which morphine acts), is an
important factor in the perception of a "pleasure landscape," and can
cancel or reduce the perception of a "pain landscape."

Let us clarify the idea a bit further with an example of pain
processing. I would say things work like this: From nerve terminals
stimulated in an area of the body where there is tissue damage (say,
the root canal in a tooth), the brain constructs a transient representation of local body change, different from the previous representation for that area. The activity pattern that corresponds to pain
signals and the perceptual characteristics of the resulting representation are prescribed entirely by the brain but otherwise are not
neurophysiologically different from any other kind of body perception. If this were all, however, I submit that all you would experience
would be a particular image of body change, without any trouble-
some consequence. You might not enjoy it, but you would not be
inconvenienced either. My point is that *the process does not stop
there*. The innocent processing of body change rapidly triggers a
wave of additional body-state changes which further deviate the
overall body state from the base range. *The state that ensues is an
emotion, with a particular profile*. It is from the subsequent body-
state deviations that the unpleasant feeling of suffering will be
formed. Why are they experienced as suffering, you may ask. Be-

cause the organism says so. We came to life with a preorganized mechanism to give us the experiences of pain and of pleasure. Culture and individual history may change the threshold at which it begins to be triggered, or its intensity, or provide us with means to dampen it. But the essential device is a given.

What is the use of having such a preorganized mechanism? Why should there be this additional state of annoyance, rather than just the pain image alone? One can only wonder, but the reason must have something to do with the fact that suffering puts us on notice. Suffering offers us the best protection for survival, since it increases the probability that individuals will heed pain signals and act to avert their source or correct their consequences.

If pain is a lever for the proper deployment of drives and instincts, and for the development of related decision-making strategies, it follows that alterations in pain perception should be accompanied by behavioral impairments. This seems to be the case. Individuals born with a bizarre condition known as congenital absence of pain do not acquire normal behavior strategies. Many seem to be eternally giggly and pleased, in spite of the fact that their condition leads to damage in their joints (deprived of pain, they move their joints well beyond the affordable mechanical limits, thus tearing ligaments and capsules), severe burns, cuts (they will not withdraw from a hot plate or a blade destroying their skin).[6] As they can still feel pleasure, and thus can be influenced by positive feelings, it is all the more interesting to find that their behavior is defective. But even more fascinating is the hypothesis that the leverage devices play a role not just in the development but also in the deployment of acquired decision-making strategies. Patients with prefrontal damage have curiously altered pain responses. Their localizable image of pain itself is intact, for example, but the emotional reactions that are part and parcel of the pain process are missing, or in the very least, the ensuing feeling is not normal. There is other evidence about this dissociation to consider, pertaining to patients in whom surgical brain lesions have been made for the treatment of chronic pain.

· · ·

Certain neurological conditions involve intense and frequent pain. One example is trigeminal neuralgia, also known as *tic douloureux*. The term neuralgia stands for pain with a neural origin, and the term trigeminal refers to the trigeminal nerve, the nerve which supplies face tissues and which ferries signals from the face to the brain. Trigeminal neuralgia affects the face, generally on one side and in one sector, for instance the cheek. Suddenly an innocent act such as touching the skin or an even more innocent breeze caressing the same skin may trigger a sudden excruciating pain. People afflicted complain of the sensation of knives' stabbing their flesh, of pins sticking in their skin and bone. Their whole lives may become focused on the pain; they can do or think of nothing else while the jabbing lasts, and the jabbing may come on frequently. Their bodies close in a tight, defensive coil.

For patients in whom the neuralgia is resistant to all available medication, the condition is classified as intractable or refractory. In such cases, neurosurgery can come to the rescue and offer the possibility of relief with a surgical intervention. One modality of treatment attempted in the past was prefrontal leucotomy (described in chapter 4). The results of this intervention illustrate better than any other fact the distinction between pain itself, that is, the perception of a certain class of sensory signals, and suffering, that is, the feeling that comes from perceiving the emotional reaction to that perception.

Consider the following episode, which I witnessed personally, when I was training with Almeida Lima, the neurosurgeon who had helped Egas Moniz develop cerebral angiography and prefrontal leucotomy and in fact had performed the first such operation. Lima, who was not only a skillful surgeon but a compassionate man, had been using a modified leucotomy for the management of intractable pain and was convinced the procedure was justifiable in desperate cases. He wanted me to see an example of the problem from the very beginning.

I vividly recall the particular patient, sitting in bed waiting for the operation. He was crouched in profound suffering, almost immobile, afraid of triggering further pain. Two days after the operation, when Lima and I visited on rounds, he was a different person. He looked relaxed, like anyone else, and was happily absorbed in a game of cards with a companion in his hospital room. Lima asked him about the pain. The man looked up and said cheerfully: "Oh, the pains are the same, but I feel fine now, thank you." Clearly, what the operation seemed to have done, then, was abolish the emotional reaction that is part of what we call pain. It had ended the man's suffering. His facial expression, his voice, and his deportment were those one associates with pleasant states, not pain. But the operation seemed to have done little to the image of local alteration in the body region supplied by the trigeminal nerve, and that is why the patient stated that the pains were the same. While the brain could no longer engender suffering, it was still making "images of pain," that is, processing normally the somatosensory mapping of a pain landscape.[7] In addition to what it may tell us about the mechanisms of pain, this example reveals the separation between the image of an entity (the state of biological tissue which equals a pain image) and the image of a body state which qualifies the entity image by dint of juxtaposition in time.

I believe that one of the main efforts of neurobiology and medicine should be directed at alleviating suffering of the sort described above. A no less important target for biomedical efforts should be the alleviation of suffering in mental diseases. But how to deal with the suffering that arises from personal and social conflicts outside the medical realm is a different and entirely unresolved matter. The current trend is to make no distinction at all and utilize the medical approach to eliminate any discomfort. The proponents of the attitude have an attractive argument. If an increase in serotonin levels, for instance, can not only treat depression but also reduce aggression, make you less shy, and turn you into a more confident person, why not take advantage of the opportunity? Would any but the most spoilsport, puritanical creature deny a fellow human being the benefits of all these wonder drugs? The problem, of course, is that the choice is not clear-cut, for a large number of reasons. First, the long-range biological effects of the drugs are unknown. Second, the consequences of socially massive drug intake are equally mysterious. Third, and perhaps most important of all: If the proposed solution to individual and social suffering bypasses the causes of individual and social conflict, it is not likely to work for very long. It may treat a symptom, but it does nothing to the roots of the disease.

I have said little about pleasure. Pain and pleasure are not twins or mirror images of each other, at least not as far as their roles in leveraging survival. Somehow, more often than not, it is the pain-related signal that steers us away from impending trouble, both at the moment and in the anticipated future. It is difficult to imagine that individuals and societies governed by the seeking of pleasure, as much as or more than by the avoidance of pain, can survive at all. Some current social developments in increasingly hedonistic cultures offer support for this opinion, and work that my colleagues and I are pursuing on the neural correlates of various emotions lends further support. There seem to be far more varieties of negative than positive emotions, and it is apparent that the brain handles positive and negative varieties of emotions with different systems. Perhaps Tolstoy had a similar insight, when he wrote, at the beginning of *Anna Karenina*: "All happy families are like one another, each unhappy family is unhappy in its own way."

Notes and References

INTRODUCTION

1. I tried to make the terms "reason," "rationality," and "decision making" as unambiguous as possible, but I must caution that their meanings are often problematic, as discussed at the beginning of Chapter 8. This is not just my problem or the reader's. A contemporary dictionary of philosophy has this to say about reason: "In English the word "reason" has long had, and still has, a large number and a wide variety of senses and uses, related to one another in ways that are often complicated and often not clear. . ." (*Encyclopedia of Philosophy*, P. Edwards, ed., 1967, New York: Macmillan Publishing Company and the Free Press.)

Be that as it may, the reader will probably find my use of the terms reason and rationality quite conventional. I generally use reason as the ability to think and make inferences in an orderly, logical manner; and rationality as the quality of thought and behavior that comes from adapting reason to a personal and social context. I do not use reasoning and decision making interchangeably since not all reasoning processes are followed by a decision.

As the reader will also discover, I do not use emotion and feeling interchangeably either. In general, I use emotion for a collection of changes occurring in both brain and body, usually prompted by a particular mental content. Feeling is the perception of those changes. A discussion of this distinction appears in Chapter 7.

2. C. Darwin (1871). *The Descent of Man*. London: Murray.

3. N. Chomsky (1984). *Modular Approaches to the Study of the Mind*. San Diego: San Diego State University Press.

4. O. Flanagan (1991). *The Science of the Mind*. Cambridge, MA: MIT Press/Bradford Books.

CHAPTER 1

1. J. M. Harlow (1868). Recovery from the passage of an iron bar through the head, *Publications of the Massachusetts Medical Society*, 2:327–47; and (1848–49). Passage of an iron rod through the head, *Boston Medical and Surgical Journal*, 39:389.

2. See note 1 above.

3. E. Williams, cited in H. J. Bigelow (1850). Dr. Harlow's case of recovery from the passage of an iron bar through the head, *American Journal of the Medical Sciences*, 19:13–22.

4. See note 3 above (Bigelow).

5. See note 1 above (1868).

6. N. West (1939). *The Day of the Locust*. Chapter 1.

7. Exemplifying this attitude is E. Dupuy (1873). *Examen de quelques points de la physiologie du cerveau*. Paris: Delahaye.

8. D. Ferrier (1878). The Goulstonian Lectures on the localisation of cerebral disease, *British Medical Journal*, 1:399–447.

9. For an exceptionally fair appraisal of Gall's contributions see J. Marshall (1980). The new organology, *The Behavioral and Brain Sciences*, 3:23–25.

10. M. B. MacMillan (1986). A wonderful journey through skull and brains, *Brain and Cognition*, 5:67–107.

11. N. Sizer (1882). *Forty Years in Phrenology; Embracing Recollec-*

tions of History, Anecdote and Experience. New York: Fowler and Wells.

12. See note 1 above (1868).

CHAPTER 2

1. P. Broca (1865). Sur la faculté du langage articulé, *Bull. Soc. Anthropol., Paris,* 6:337–93.

C. Wernicke (1874). *Der aphasische Symptomencomplex.* Breslau: Cohn und Weigert.

For details on Broca and Wernicke aphasias, see A. Damasio (1992). *The New England Journal of Medicine,* 326:531–39. For a recent view on the neuroanatomy of language, see A. Damasio and H. Damasio (1992). *Scientific American,* 267: 89–95.

2. For a general text on neuroanatomy, see J. H. Martin (1989). *Neuroanatomy Text and Atlas.* New York: Elsevier. For a modern atlas of the human brain, see H. Damasio (1994). *Human Neuroanatomy from Computerized Images.* New York: Oxford University Press. For a comment on the importance of neuroanatomy in the future of neurobiology, see F. Crick and E. Jones (1993). The Backwardness of human neuroanatomy, *Nature,* 361:109–10.

3. H. Damasio and R. Frank (1992). Three-dimensional *in vivo* mapping of brain lesions in humans, *Archives of Neurology,* 49:137–43.

4. See E. Kandel, J. Schwartz, T. Jessell (1991). *Principles of Neuroscience.* Amsterdam: Elsevier.

P. S. Churchland and T. J. Sejnowski (1992). *The Computational Brain: Models and Methods on the Frontiers of Computational Neuroscience.* Boston: MIT Press, Bradford Books.

5. H. Damasio, T. Grabowski, R. Frank, A. M. Galaburda, and A. R. Damasio (1994). The return of Phineas Gage: The skull of a famous patient yields clues about the brain, *Science,* 264: 1102–05.

CHAPTER 3

1. With the exception of Phineas Gage, the privacy of all patients mentioned in the text is protected by coded initials, pen names, and by omission of identifying biographic details.

2. Many of the neuropsychological tests to which I refer in this section are described in M. Lezak (1983). *Neuropsychological Assessment*. New York: Oxford University Press; and A. L. Benton (1983). *Contributions to Neuropsychological Assessment*. New York: Oxford University Press.

3. B. Milner (1964). Some effects of frontal lobectomy in man, in J. M. Warren and K. Akert, eds., *The Frontal Granular Cortex and Behavior*. New York: McGraw-Hill.

4. T. Shallice and M. E. Evans (1978). The involvement of the frontal lobes in cognitive estimation, *Cortex*, 14:294–303.

5. S. R. Hathaway and J. C. McKinley (1951). *The Minnesota Multiphasic Personality Inventory Manual* (rev. ed.). New York: Psychological Corporation.

6. L. Kohlberg (1987). *The Measurement of Moral Judgment*. Cambridge, Massachusetts: Cambridge University Press.

7. J. L. Saver and A. R. Damasio (1991). Preserved access and processing of social knowledge in a patient with acquired sociopathy due to ventromedial frontal damage, *Neuropsychologia*, 29: 1241–49.

CHAPTER 4

1. B. J. McNeil, S. G. Pauker, H. C. Sox, and A. Tversky (1982). On the elicitation of preferences for alternative therapies, *New England Journal of Medicine*, 306:1259–69.

2. For details on neuropsychology research strategy, see H. Damasio and A. R. Damasio (1989). *Lesion Analysis in Neuropsychology*. New York: Oxford University Press.

3. R. M. Brickner (1934). An interpretation of frontal lobe function based upon the study of a case of partial bilateral frontal lobectomy,

Research Publications of the Association for Research in Nervous and Mental Disease, 13:259–351; and (1936). *The intellectual functions of the frontal lobes: Study based upon observation of a man after partial bilateral frontal lobectomy.* New York: Macmillan. For other studies of frontal lobe damage, see also D. T. Stuss and F. T. Benson (1986). *The Frontal Lobes.* New York: Raven Press.

4. D. O. Hebb and W. Penfield (1940). Human behavior after extensive bilateral removals from the frontal lobes, *Archives of Neurology and Psychiatry*, 44:421–38.

5. S. S. Ackerly and A. L. Benton (1948). Report of a case of bilateral frontal lobe defect, *Research Publications of the Association for Research in Nervous and Mental Disease*, 27:479–504.

6. Among the few documentations of cases comparable to that of Ackerly and Benton's patient are the following:

B. H. Price, K. R. Daffner, R. M. Stowe, and M. M. Mesulam (1990). The comportmental learning disabilities of early frontal lobe damage, *Brain*, 113:1383–93.

L. M. Grattan, and P. J. Eslinger (1992). Long-term psychological consequences of childhood frontal lobe lesion in patient DT, *Brain and Cognition*, 20:185–95.

7. E. Moniz (1936). *Tentatives opératoires dans le traitement de certaines psychoses.* Paris: Masson.

8. For a discussion on this and other forms of aggressive treatment see E. S. Valenstein (1986). *Great and Desperate Cures: The Rise and Decline of Psychosurgery and Other Radical Treatment for Mental Illness.* New York: Basic Books.

9. J. Babinski (1914). Contributions à l'étude des troubles mentaux dans l'hémiplégie organique cérébrale (anosognosie), *Revue neurologique*, 27:845–47.

10. A. Marcel (1993). Slippage in the unity of consciousness, in *Experimental and theoretical studies of consciousness* (Ciba Foundation Symposium 174), pp. 168–86. New York: John Wiley & Sons.

11. S. W. Anderson and D. Tranel (1989). Awareness of disease states following cerebral infarction, dementia, and head trauma: Standardized assessment, *The Clinical Neuropsychologist*, 3:327–39.

12. R. W. Sperry (1981). Cerebral organization and behavior, *Science*, 133:1749–57.

J. E. Bogen and G. M. Bogen (1969). The other side of the brain. III: The corpus callosum and creativity, *Bull. Los Angeles Neurol. Soc.*, 34:191–220.

E. De Renzi (1982). *Disorders of Space Exploration and Cognition*. New York: John Wiley & Sons.

D. Bowers, R. M. Bauer, and K. M. Heilman (1993). The nonverbal affect lexicon: Theoretical perspectives from neuropsychological studies of affect perception, *Neuropsychologia*, 7:433–44.

M. M. Mesulam (1981). A cortical network for directed attention and unilateral neglect, *Ann. Neurol.*, 10:309–25.

E. D. Ross and M. M. Mesulam (1979). Dominant language functions of the right hemisphere, *Arch. Neurol.*, 36:144–48.

13. B. Woodward and S. Armstrong (1979). *The Brethren*. New York: Simon & Schuster.

14. D. Tranel and B. T. Hyman (1990). Neuropsychological correlates of bilateral amygdala damage, *Archives of Neurology*, 47:349–55.

F. K. D. Nahm, H. Damasio, D. Tranel, and A. Damasio (1993). Cross-modal associations and the human amygdala, *Neuropsychologia*, 31:727–44.

R. Adolphs, D. Tranel, and A. Damasio. Bilateral Damage to the Human Amygdala Impairs the Recognition of Emotion in Facial Expressions. (to appear)

15. L. Weiskrantz (1956). Behavioral changes associated with ablations of the amygdaloid complex in monkeys, *Journal of Comparative and Physiological Psychology*, 49:381–91.

J. P. Aggleton and R. E. Passingham (1981). Syndrome produced by lesions of the amygdala in monkeys (*Macaca mulatta*), *Journal of Comparative and Physiological Psychology*, 95:961–77.

For studies on rats, see J. E. LeDoux (1992). Emotion and the amygdala, in J. P. Aggleton, ed., *The Amygdala: Neurobiological Aspects of Emotion, Mystery, and Mental Dysfunction*, pp. 339–51. New York: Wiley-Liss.

16. R. J. Morecraft and G. W. Van Hoesen (1993). Frontal granular cortex input to the cingulate (M3), supplementary (M2), and primary (M1) motor cortices in the rhesus monkey, *Journal of Comparative Neurology,* 337:669–89.

17. A. R. Damasio and G. W. Van Hoesen (1983). Emotional disturbances associated with focal lesions of the limbic frontal lobe, in K. M. Heilman and P. Satz, eds., *Neuropsychology of Human Emotion.* New York: The Guilford Press.

M. I. Posner and S. E. Petersen (1990). The attention system of the human brain, *Annual Review of Neuroscience,* 13:25–42.

18. F. Crick (1994). *The Astonishing Hypothesis: The Scientific Search for the Soul.* New York: Charles Scribner's Sons.

19. J. F. Fulton and C. F. Jacobsen (1935). The functions of the frontal lobes: A comparative study in monkeys, chimpanzees and man, *Advances in Modern Biology (Moscow),* 4:113–23.

J. F. Fulton (1951). *Frontal Lobotomy and Affective Behavior.* New York: Norton and Company.

20. C. F. Jacobsen (1935). Functions of the frontal association area in primates, *Archives of Neurology and Psychiatry,* 33:558–69.

21. R. E. Myers (1975). Neurology of social behavior and affect in primates: A study of prefrontal and anterior temporal cortex, in K. J. Zuelch, O. Creutzfeldt, and G. C. Galbraith, eds., *Cerebral Localization,* pp. 161–70. New York: Springer-Verlag.

E. A. Franzen and R. E. Myers (1973). Neural control of social behavior: Prefrontal and anterior temporal cortex, *Neuropsychologia,* 11:141–57.

22. S. J. Suomi (1987). Genetic and maternal contributions to individual differences in rhesus monkey biobehavioral development. In *Perinatal Development: A Psychobiological Perspective,* pp. 397–419. New York: Academic Press, Inc.

23. For a review of neurophysiological evidence on this issue, see L. Brothers, Neurophysiology of social interactions, in M. Gazzaniga, ed., *The Cognitive Neurosciences* (in press).

24. P. Goldman-Rakic (1987). Circuitry of primate prefrontal cortex and regulation of behavior by representational memory, in F. Plum

and V. Mountcastle, eds., *Handbook of Physiology: The Nervous System*, vol. 5, pp. 373–417. Bethesda, MD: American Physiological Society.

J. M. Fuster (1989). *The Prefrontal Cortex: Anatomy, Physiology, and Neuropsychology of the Frontal Lobe* (2nd ed.). New York: Raven Press.

25. M. J. Raleigh and G. L. Brammer (1993). Individual differences in serotonin-2 receptors and social behavior in monkeys, *Society for Neuroscience Abstracts*, 19:592.

CHAPTER 5

1. E. G. Jones and T. P. S. Powell (1970). An anatomical study of converging sensory pathways within the cerebral cortex of the monkey, *Brain*, 93:793–820. The work of the neuroanatomists D. Pandya, K. Rockland, G. W. Van Hoesen, P. Goldman-Rakic, and D. Van Essen has repeatedly confirmed this connectional principle and elucidated its intricacies.

2. D. Dennett (1991). *Consciousness Explained*. Boston: Little, Brown.

3. A. R. Damasio (1989). The brain binds entities and events by multiregional activation from convergence zones, *Neural Computation*, 1:123–32.

——— (1989). Time-locked multiregional retroactivation: A systems level proposal for the neural substrates of recall and recognition, *Cognition*, 33:25–62.

A. R. Damasio and H. Damasio (1993). Cortical systems underlying knowledge retrieval: Evidence from human lesion studies, in *Exploring Brain Functions: Models in Neuroscience*, pp. 233–48. New York: Wiley & Sons.

——— (1994). Cortical systems for retrieval of concrete knowledge: The convergence zone framework, in C. Koch, ed., *Large-Scale Neuronal Theories of the Brain*. Cambridge, MA: MIT Press.

4. Among others, see:

C. von der Malsburg (1987). Synaptic plasticity as basis of brain organization, in P.-P. Changeux and M. Konishi, eds., *The Neural and Molecular Bases of Learning* (Dahlem Workshop Report 38), pp. 411–31. Chichester, England: Wiley.

G. Edelman (1987). *Neural Darwinism: The Theory of Neuronal Group Selection.* New York: Basic Books.

R. Llinas (1993). Coherent 40-Hz oscillation characterizes dream state in humans, *Proceedings of the National Academy of Sciences,* 90:2078–81.

F. H. Crick and C. Koch (1990). Towards a neurobiological theory of consciousness, *Seminars in the Neurosciences,* 2:263–75.

W. Singer, A. Artola, A. K. Engel, P. Koenig, A. K. Kreiter, S. Lowel, and T. B. Schillen (1993). Neuronal representations and temporal codes, in T. A. Poggio and D. A. Glaser, eds., *Exploring Brain Functions: Models in Neuroscience,* pp. 179–94. Chichester, England: Wiley.

R. Eckhorn, R. Bauer, W. Jordan, M. Brosch, W. Kruse, M. Munk, and H. J. Reitboeck (1988). Coherent oscillations: A mechanism for feature linking in the visual cortex, *Biologica Cybernetica,* 60:121–30.

S. Zeki (1993). *A Vision of the Brain.* London: Blackwell Scientific.

S. Bressler, R. Coppola, and R. Nakamura (1993). Episodic multi-regional cortical coherence at multiple frequencies during visual task performance, *Nature,* 366:153–56.

5. See the discussion in chapter 4 of this book, and see: M. I. Posner and S. E. Petersen (1990). The attention system of the human brain, *Annual Review of Neuroscience,* 13:25–42. P. S. Goldman-Rakic (1987). Circuitry of primate prefrontal cortex and regulation of behavior by representational memory, in F. Plum and V. Mountcastle, eds., *Handbook of Physiology: The Nervous System,* vol. 5, pp. 373–417. Bethesda, MD: American Physiological Society.

J. M. Fuster (1989). *The Prefrontal Cortex: Anatomy, Physiology, and Neuropsychology of the Frontal Lobe* (2nd ed.). New York: Raven Press.

6. For neuroanatomical, neurophysiological, and psychophysical studies concerning vision, see:

J. Allman, F. Miezin, and E. McGuiness (1985). Stimulus specific

responses from beyond the classical receptive field: Neuropsychological mechanisms for local-global comparisons in visual neurons,
Annual Review of Neuroscience, 8:407–30.

W. Singer, C. Gray, A. Engel, P. Koenig, A. Artola, and S. Brocher
(1990). Formation of cortical cell assemblies, *Symposia on Quantitative Biology*, 55:939–52.

G. Tononi, O. Sporns, and G. Edelman (1992). Reentry and the
problem of integrating multiple cortical areas: Simulation of dynamic integration in the visual system, *Cerebral Cortex*, 2:310–35.

S. Zeki (1992). The visual image in mind and brain, *Scientific American*, 267:68–76.

For the somatosensory and auditory studies, see:

R. Adolphs (1993). Bilateral inhibition generates neuronal responses
tuned to interaural level differences in the auditory brainstem of the
barn owl, *The Journal of Neuroscience*, 13:3647–68.

M. Konishi, T. Takahashi, H. Wagner, W. E. Sullivan, and C. E. Carr
(1988). Neurophysiological and anatomical substrates of sound localization in the owl, in G. Edelman, W. Gall, and W. Cowan, eds.,
Auditory Function, pp. 721–46. New York: John Wiley & Sons.

M. M. Merzenich and J. H. Kaas (1980). Principles of organization
of sensory-perceptual systems in mammals, in J. M. Sprague and
A. N. Epstein, eds., *Progress in Psychobiology and Physiological Psychology*, pp. 1–42. New York: Academic Press.

For studies on cortical plasticity, see:

C. D. Gilbert, J. A. Hirsch, and T. N. Wiesel (1990). Lateral interactions in visual cortex. In: *Symposia on Quantitative Biology*, vol. 55,
pp. 663–77. Cold Spring Harbor, N.Y.: Laboratory Press.

M. M. Merzenich, J. H. Kaas, J. Wall, R. J. Nelson, M. Sur, and D.
Felleman (1983). Topographic reorganization of somatosensory cortical areas 3B and 1 in adult monkeys following restructured deaf-
ferentation, *Neuroscience*, 8:33–55.

V. S. Ramachandran (1993). Behavioral and magnetoencephalo-
graphic correlates of plasticity in the adult human brain, *Proceedings
of the National Academy of Science*, 90:10413–20.

7. F. C. Bartlett (1964). *Remembering: A Study in Experimental and*

Social Psychology. Cambridge, England: Cambridge University Press.

8. S. M. Kosslyn, N. M. Alpert, W. L. Thompson, V. Maljkovic, S. B. Weise, C. F. Chabris, S. E. Hamilton, S. L. Rauch, and F. S. Buonanno (1993). Visual mental imagery activates topographically organized visual cortex: PET investigations, *Journal of Cognitive Neuroscience,* 5:263–87.

H. Damasio, T. J. Grabowski, A. Damasio, D. Tranel, L. Boles-Ponto, G. L. Watkins, and R. D. Hichwa (1993). Visual recall with eyes closed and covered activates early visual cortices, *Society for Neuroscience Abstracts,* 19:1603.

9. The pathways for the backfiring are beginning to be understood. See:

G. W. Van Hoesen (1982). The parahippocampal gyrus: New observations regarding its cortical connections in the monkey, *Trends in Neurosciences,* 5:345–50.

M. S. Livingstone and D. H. Hubel (1984). Anatomy and physiology of a color system in the primate visual cortex, *The Journal of Neuroscience,* 4:309–56.

D. H. Hubel and M. S. Livingstone (1987). Segregation of form, color, and stereopsis in primate area 18, *The Journal of Neuroscience,* 7:3378–3415.

M. S. Livingstone and D. H. Hubel (1987). Connections between layer 4B of area 17 and thick cytochrome oxidase stripes of area 18 in the squirrel monkey, *The Journal of Neuroscience,* 7:3371–77.

K. S. Rockland and A. Virga (1989). Terminal arbors of individual "feedback" axons projecting from area V2 to V1 in the macaque monkey: A study using immunohistochemistry of anterogradely transported *Phaseolus vulgaris* leucoagglutinin, *Journal of Comparative Neurology,* 285:54–72.

D. J. Felleman and D. C. Van Essen (1991). Distributed hierarchical processing in the primate cerebral cortex, *Cerebral Cortex,* 1:1–47.

10. R. B. H. Tootell, E. Switkes, M. S. Silverman, and S. L. Hamilton (1988). Functional anatomy of macaque striate cortex. II. Retinotopic organization, *The Journal of Neuroscience,* 8:1531–68.

11. M. M. Merzenich, note 3 above.

12. It is not possible to do justice here to the scientific literature on learning and plasticity. The reader is referred to selected chapters in two books:

E. Kandel, J. Schwartz, and T. Jessell (1991). *Principles of Neuroscience.* Amsterdam: Elsevier.

P. S. Churchland and T. J. Sejnowski (1992). *The Computational Brain: Models and Methods on the Frontiers of Computational Neuroscience.* Cambridge, MA: MIT Press/Bradford Books.

13. The value accorded to images is a recent development, part of the cognitive revolution that followed the long night of stimulus-response behaviorism. We owe it in large part to the work of Roger Shepard and Stephen Kosslyn. See:

R. N. Shepard and L. A. Cooper (1982). *Mental Images and Their Transformations.* Cambridge, MA: MIT Press.

S. M. Kosslyn (1980). *Image and Mind.* Cambridge, MA: Harvard University Press.

For a historical review, see also Howard Gardner (1985). *The Mind's New Science.* New York: Basic Books.

14. B. Mandelbrot, personal communication.

15. A. Einstein, cited in J. Hadamard (1945). *The Psychology of Invention in the Mathematical Field.* Princeton, NJ: Princeton University Press.

16. The following are key references on this subject: D. H. Hubel and T. N. Wiesel (1965). Binocular interaction in striate cortex of kittens reared with artificial squint, *Journal of Neurophysiology,* 28:1041–59.

D. H. Hubel, T. N. Wiesel, and S. LeVay (1977). Plasticity of ocular dominance columns in monkey striate cortex, *Philosophical Transactions of the Research Society of London,* ser. B, 278:377–409.

L. C. Katz and M. Constantine-Paton (1988). Relationship between segregated afferents and post-synaptic neurons in the optic tectum of three-eyed frogs, *The Journal of Neuroscience,* 8:3160–80.

G. Edelman (1988). *Topobiology.* New York: Basic Books.

M. Constantine-Paton, H. T. Cline, and E. Debski (1990). Patterned activity, synaptic convergence, and the NMDA receptor in developing visual pathways, *Annual Review of Neuroscience*, 13:129–54. C. Shatz (1992). The developing brain, *Scientific American*, 267:61–67.

17. For pertinent background on this issue see: R. C. Lewontin (1992). *Biology as Ideology*. New York: Harper Perennial; Stuart A. Kauffman (1993). *The Origins of Order. Self-Organization and Selection in Evolution*. New York: Oxford University Press.

18. The substrate of the rapid and dramatic changes in circuit design that seem to occur, include the wealth of synapses to which I previously alluded, enriched by the variety of neurotransmitters and receptors available at each synapse. The characterization of this plastic process is outside the scope of this text, but the account provided here is compatible with the idea that it largely occurs by selection of circuitries at synaptic level. The application of the notion of selection to the nervous system was first suggested by Niels Jerne and J. Z. Young and used by Jean Pierre Changeux. Gerald Edelman has championed the idea and built a comprehensive theory of mind and brain around it.

CHAPTER 6

1. C. B. Pert, M. R. Ruff, R. J. Weber, and M. Herkenham (1985). Neuropeptides and their receptors: A psychosomatic network, *The Journal of Immunology*, 135:820s–26s.

F. Bloom (1985). Neuropeptides and other mediators in the central nervous system, *The Journal of Immunology*, 135:743s–45s.

J. Roth, D. LeRoith, E. S. Collier, N. R. Weaver, A. Watkinson, C. F. Cleland, and S. M. Glick (1985). Evolutionary origins of neuropeptides, hormones, and receptors: Possible applications to immunology, *The Journal of Immunology*, 135:816s–19s.

B. S. McEwen (1991). Non-genomic and genomic effects of steroids on neural activity, *Trends in Pharmacological Sciences*, Apr:12(4): 141–7.

A. Herzog (1984). Temporal lobe epilepsy: An extrahypothalamic pathogenesis for polycystic ovarian syndrome?, *Neurology*, 34:1389–93.

2. J. Hosoi, G. F. Murphy, and C. L. Egan (1993). Regulation of

Langerhans cell function by nerves containing calcitonin gene-related peptide, *Nature,* 363:159–63.

3. J. R. Calabrese, M. A. Kling, and P. Gold (1987). Alterations in immunocompetence during stress, bereavement and depression: Focus on neuroendocrine regulation, *American Journal of Psychiatry,* 144:1123–34.

4. E. Marder, ed. (1989). Neuromodulation in circuits underlying behavior, *Seminars in the Neurosciences,* 1:3–4.

C. B. Saper (1987). Diffuse cortical projection systems: anatomical organization and role in cortical function. In: V. B. Mountcastle, ed., *Handbook of Physiology,* pp. 169–210. Bethesda, Maryland: American Physiological Society.

5. C. S. Carter (1992). Oxytocin and sexual behavior, *Neuroscience Biobehavioral Review,* 16:131.

T. R. Insel (1992). Oxytocin, a neuropeptide for affiliation: Evidence from behavioral, receptor autoradiographic, and comparative studies, *Psychoneuroendocrinology,* 17:3.

6. R. Descartes (1647). *The Passions of the Soul,* in J. Cottingham, R. Stoothoff, and D. Murdoch, eds., *The Philosophical Writings of Descartes,* vol. 1. Cambridge, England: Cambridge University Press (1985).

7. S. Freud (1930). *Civilization and Its Discontents.* Chicago: University of Chicago Press.

CHAPTER 7

1. J. M. Allman, T. McLaughlin, and A. Hakeem (1993). Brain weight and life-span in primate species, *Proceedings of the National Academy of Science,* 90:118–22.

2. ———. Brain structures and life-span in primate species, *Proceedings of the National Academy of Science,* 90:3559–63.

3. W. James (1890). *The Principles of Psychology,* vol. 2. New York: Dover (1950).

4. As an introduction to the extensive scholarship on the subject, I recommend the following:

P. Ekman (1992). Facial expressions of emotion: New findings, new questions, *Psychological Science,* 3:34–38.

R. S. Lazarus (1984). On the primacy of cognition, *American Psychologist,* 39:124–29.

G. Mandler (1984). *Mind and Body: Psychology of Emotion and Stress.* New York: W. W. Norton & Co.

R. B. Zajonc (1984). On the primacy of affect, *American Psychologist,* 39:117–23.

5. M. H. Bagshaw, D. P. Kimble, and K. H. Pribram (1965). The GSR of monkeys during orienting and habituation and after ablation of the amygdala, hippocampus and inferotemporal cortex, *Neuropsychologia,* 3:111–19.

L. Weiskrantz (1956). Behavioral changes associated with ablations of the amygdaloid complex in monkeys, *Journal of Comparative and Physiological Psychology,* 49:381–91.

J. P. Aggleton and R. E. Passingham (1981). Syndrome produced by lesions of the amygdala in monkeys (*Macaca mulatta*), *Journal of Comparative and Physiological Psychology,* 95:961–77.

J. E. LeDoux (1992). Emotion and the amygdala, in J. P. Aggleton, ed., *The Amygdala: Neurobiological Aspects of Emotion, Memory, and Mental Dysfunction,* pp. 339–51. New York: Wiley-Liss.

6. M. Davis (1992). The role of the amygdala in conditioned fear, in J. P. Aggleton, ed., *The Amygdala: Neurobiological Aspects of Emotion, Memory, and Mental Dysfunction,* pp. 255–305. New York: Wiley-Liss.

S. Zola-Morgan, L. R. Squire, P. Alvarez-Royo, and R. P. Clower (1991). Independence of memory functions and emotional behavior: Separate contributions of the hippocampal formation and the amygdala, *Hippocampus,* 1:207–20.

7. P. Gloor, A. Olivier, and L. F. Quesney (1981). The role of the amygdala in the expression of psychic phenomena in temporal lobe seizures, in Y. Ben-Air, ed., *The Amygdaloid Complex* (INSERM Symposium 20), pp. 489–98. Amsterdam: Elsevier North-Holland.

W. Penfield and W. Jasper (1954). *Epilepsy and the Functional Anatomy of the Human Brain.* Boston: Little, Brown.

8. H. Kluver and P. C. Bucy (1937). "Psychic blindness" and other symptoms following bilateral temporal lobe lobectomy in rhesus monkeys, *American Journal of Physiology*, 119:352–53.

9. D. Laplane, J. D. Degos, M. Baulac, and F. Gray (1981). Bilateral infarction of the anterior cingulate gyri and of the fornices, *Journal of the Neurological Sciences*, 51:289–300; and A. R. Damasio and G. W. Van Hoesen (1983). Emotional disturbances associated with focal lesions of the limbic frontal lobe, in K. M. Heilman and P. Satz, eds., *Neuropsychology of Human Emotion*. New York: The Guilford Press.

10. R. W. Sperry, M. S. Gazzaniga, and J. E. Bogen (1969). Interhemispheric relationships: The neocortical commissures; syndromes of their disconnection, in P. J. Vinken and G. W. Bruyn, eds., *Handbook of Clinical Neurology*, vol. 4, pp. 273–90. Amsterdam: North Holland; R. Sperry, E. Zaidel, and D. Zaidel (1979). Self recognition and social awareness in the deconnected minor hemisphere, *Neuropsychologia*, 17:153–66.

11. G. Gainotti (1972). Emotional behavior and hemispheric side of the lesion, *Cortex*, 8:41–55.

H. Gardner, H. H. Brownell, W. Wapner, and D. Michelow (1983). Missing the point: The role of the right hemisphere in the processing of complex linguistic materials, in E. Pericman, ed., *Cognitive Processes and the Right Hemisphere*. New York: Academic Press.

K. Heilman, R. T. Watson, and D. Bowers (1983). Affective disorders associated with hemispheric disease, in K. Heilman and P. Satz, eds., *Neuropsychology of Human Emotion*, pp. 45–64. New York: The Guilford Press.

J. C. Borod (1992). Interhemispheric and intrahemispheric control of emotion: A focus on unilateral brain damage, *Journal of Consulting and Clinical Psychology*, 60:339–48.

R. Davidson (1992). Prolegomenon to emotion: Gleanings from Neuropsychology, *Cognition and Emotion*, 6:245–68.

12. C. Darwin (1872). *The Expression of the Emotions in Man and Animals*. New York: Philosophical Library.

13. G.-B. Duchenne (1862). *The Mechanism of Human Facial Ex-*

pression, or An Electro-Physiological Analysis of the Expression of the Emotions, trans. R. A. Cuthberton. New York: Cambridge University Press (1990).

14. P. Ekman (1992). Facial expressions of emotion: New findings, new questions, *Psychological Science,* 3:34–38.

P. Ekman and R. J. Davidson (1993). Voluntary smiling changes regional brain activity, *Psychological Science,* 4:342–45.

P. Ekman, R. W. Levenson, and W. V. Friesen (1983). Autonomic nervous system activity distinguishes among emotions, *Science,* 221:1208–10.

15. P. Ekman and R. J. Davidson (1993). Voluntary smiling changes regional brain activity, *Psychological Science,* 4:342–45.

16. While there appears to be a large biological component to what I have called primary emotions, the way we conceptualize secondary emotions is relative to particular cultures (for evidence on how culture contributes to the way we categorize emotions, see James A. Russell [1991]. Culture and the Categorization of Emotions, *Psychological Bulletin,* 110:426–50).

17. O. Sacks (1987). *The Man Who Mistook His Wife for a Hat, and Other Clinical Tales.* New York: Harper & Row. Part I. Chapter 3, pg. 43.

18. William Styron's memoir can be offered again as an insightful illustration of these many lines of operation. Some evidence for the picture I am drawing here can also be gleaned from studies of conceptual style in writers. N. J. Andreasen and P. S. Powers (1974). Creativity and psychosis: An examination of conceptual style, *Archives of General Psychiatry,* 32:70–73.

CHAPTER 8

1. Blaise Pascal. *Pensées.* (1670). The source used for this book was the "new edition" published by Mercure de France, 1976, Paris. The passage cited on page 165 appears under section 80.

"Que chacun examine ses penseés, il les trouvera toutes occupées au passé ou à l'avenir. Nous ne pensons presque point au présent, et si nous y pensons, ce n'est que pour en prendre la lumière pour disposer de l'avenir."

The passage cited on page 200 appears under section 680.

"Le coeur a ses raisons, que la raison ne connaît point." Author translations.

2. Phillip N. Johnson-Laird and Eldar Shafir (1993). The interaction between reasoning and decision-making: an introduction, *Cognition*, 49:109.

3. H. Gardner (1983). *Frames of Mind: The Theory of Multiple Intelligences*. New York: Basic Books.

4. A. Tversky and D. Kahneman (1973). Availability: A heuristic for judging frequency and probability, *Cognitive Psychology*, 2:207–32.

5. S. Sutherland (1992). *Irrationality: The Enemy Within*. London: Constable.

6. L. Cosmides (1989). The logic of social exchange: Has natural selection shaped how humans reason? Studies with the Wason selection task, *Cognition*, 33:187–276.

Jerome H. Barkow, Leda Cosmides, and John Tooby (eds.), *The Adapted Mind: Evolutionary Psychology and the Generation of Culture*. New York: Oxford University Press (1992).

L. Brothers, ch. 4, note 23, and Suomi, ch. 4, note 22.

7. On frontal anatomy, see F. Sanides (1964). The cytomyeloarchitecture of the human frontal lobe and its relation to phylogenetic differentiation of the cerebral cortex, *Journal für Hirnforschung*, 6:269–82.

P. Goldman-Rakic (1987). Circuitry of primate prefrontal cortex and regulation of behavior by representational memory, in F. Plum and V. Mountcastle, eds., *Handbook of Physiology: The Nervous System*, vol. 5, pp. 373–401. Bethesda, MD: American Physiological Society.

D. Pandya and E. H. Yeterian (1990). Prefrontal cortex in relation to other cortical areas in rhesus monkey: architecture and connections, in H. B. M. Uylings, ed., *The Prefrontal Cortex: Its Structure, Function and Pathology*, pp. 63–94. Amsterdam: Elsevier.

H. Barbas and D. N. Pandya (1989). Architecture and intrinsic connections of the prefrontal cortex in the rhesus monkey, *The Journal of Comparative Neurology*, 286:353–75.

8. M. Petrides and B. Milner (1982). Deficits on subject-ordered tasks after frontal and temporal lobe lesions in man, *Neuropsychologia* 20:249–62.

J. M. Fuster (1989). *The Prefrontal Cortex: Anatomy, Physiology, and Neuropsychology of the Frontal Lobe* (2nd ed.). New York: Raven Press.

P. Goldman-Rakic (1992). Working memory and the mind, *Scientific American,* 267:110–17.

9. R. J. Morecraft and G. W. Van Hoesen (1993). Frontal granular cortex input to the cingulate (M3), supplementary (M2), and primary (M1) motor cortices in the rhesus monkey, *Journal of Comparative Neurology,* 337:669–89.

10. L. A. Real (1991). Animal choice behavior and the evolution of cognitive architecture, *Science,* 253:980–86.

11. P. R. Montague, P. Dayan, and T. J. Sejnowski (1993). Foraging in an uncertain world using predictive hebbian learning, *Society for Neuroscience,* 19:1609.

12. H. Poincaré (1908). Le raisonnement mathématique, in *Science et méthode.* Translation by George Bruce Halsted, in B. Chiselin, *The Creative Process.* Los Angeles: Mentor Books/UCLA (1955).

13. L. Szilard in W. Lanouette, *Genius in the Shadows.* New York: Charles Scribner's Sons (1992).

14. J. Salk (1985). *The Anatomy of Reality.* New York: Praeger.

15. T. Shallice and P. W. Burgess (1993). Supervisory control of action and thought selection. In *Attention: Selection, Awareness, and Control: A Tribute to Donald Broadbent,* A. Baddeley and L. Weiskrantz (eds.). Oxford: Clarendon Press, pp. 171–87.

16. See note 4 above.

17. See note 5 above.

18. G. Harrer and H. Harrer (1977). Music, emotion and autonomic function, in M. Critchley and R. A. Henson, eds., *Music and the Brain,* pp. 202–215. London: William Heinemann Medical.

19. S. Dehaene and J.-P. Changeux (1991). The Wisconsin Card

Sorting Test: Theoretical analysis and modeling in a neuronal network, *Cerebral Cortex*, 1:62–79.

20. See Posner and Petersen, ch. 4, note 17.

21. See Goldman-Rakic, Working Memory and the Mind, ch. 8, note 7.

22. K. S. Lashley (1951). The problem of serial order in behavior, in L. A. Jeffress, ed., *Cerebral Mechanisms in Behavior*. New York: John Wiley & Sons.

23. C. D. Salzman, and W. T. Newsome (1994). Neural mechanisms for forming a perceptual decision, *Science*, 264:231–37.

24. Blaise Pascal (1670). *Pensées*. See note 1 above.

25. J. St. B. T. Evans, D. E. Over, and K. I. Manktelow (1993). Reasoning, decision-making and rationality, *Cognition*, 49:165–87. R. De Sousa (1991). *The Rationality of Emotion*. Cambridge, MA: MIT Press.

P. N. Johnson-Laird, and K. Oatley (1992). Basic emotions, rationality, and folk theory, *Cognition and Emotion*, 6:201–23.

CHAPTER 9

1. A. R. Damasio, D. Tranel, and H. Damasio (1991). Somatic markers and the guidance of behavior: Theory and preliminary testing, in H. S. Levin, H. M. Eisenberg, and A. L. Benton, eds., *Frontal Lobe Function and Dysfunction*, pp. 217–29. New York: Oxford University Press.

It is interesting to note that, in very similar experiments, individuals diagnosed with developmental psychopathy and with a criminal record behaved quite similarly. See R. D. Hare and M. J. Quinn (1971). Psychopathy and autonomic conditioning, *Journal of Abnormal Psychology*, 77:223–35.

2. A. Bechara, A. R. Damasio, H. Damasio, and S. Anderson (1994). Insensitivity to future consequences following damage to human prefrontal cortex, *Cognition*, 50:7–12.

3. C. M. Steele and R. A. Josephs (1990). Alcohol myopia, *American Psychologist*, 45:921–33.

4. A. Bechara, D. Tranel, H. Damásio, and A. R. Damasio (1993). Failure to respond autonomically in anticipation of future outcomes following damage to human prefrontal cortex, *Society for Neuroscience*, 19:791. Full article to appear 1994.

CHAPTER 10

1. G. Lakoff (1987). *Women, Fire, and Dangerous Things: What Categories Reveal About the Mind*. Chicago: University of Chicago Press.
M. Johnson (1987). *The Body in the Mind: The Bodily Basis of Meaning, Imagination, and Reason*. Chicago: University of Chicago Press.
2. G. W. Hohmann (1966). Some effects of spinal cord lesions on experienced emotional feelings, *Psychophysiology*, 3:143–56.
3. H. Putnam (1981). *Reason, Truth, and History*. Cambridge, England: Cambridge University Press.
4. For a review on visceral aspects of somatosensory representation, see M. M. Mesulam and E. J. Mufson (1985). The insula of Reil in man and monkey. In A. Peters and E. G. Jones (eds.): *Cerebral Cortex*, Vol. 5. New York, Plenum Press, pp. 179–226. Also see J. R. Jennings (1992). Is it important that the mind is in the body? Inhibition and the heart, *Psychophysiology*, 29:369–83. See also S. M. Oppenheimer, A. Gelb, J. P. Girvin, and V. C. Hachinski (1992). Cardiovascular effects of human insular cortex stimulation, *Neurology*, 42:1727–32.
5. N. Humphrey (1992). *A History of the Mind*. New York: Simon & Schuster.
6. See note 1 above, and
F. Varela, E. Thompson, and E. Rosch (1992). *The Embodied Mind*. Cambridge, MA: MIT Press.
G. Edelman (1992). *Bright Air, Brilliant Fire*. New York: Basic Books.
7. J. Searle (1992). *The Rediscovery of the Mind*. Cambridge, MA: MIT Press.
P. S. Churchland (1986). *Neurophilosophy: Toward a Unified*

Science of the Mind-Brain. Cambridge, MA: Bradford Books/MIT Press.

P. M. Churchland (1984). *Matter and Consciousness*. Cambridge, MA: Bradford Books/MIT Press.

F. Crick (1994). *The Astonishing Hypothesis: The Scientific Search for the Soul*. New York: Charles Scribner's Sons.

D. Dennett (1991). *Consciousness Explained*. Boston: Little, Brown.

G. Edelman, see note 6 above.

R. Llinás (1991). Commentary of dreaming and wakefulness, *Neuroscience*, 44:521–35.

8. F. Plum and J. Posner (1980). *The Diagnosis of Stupor and Coma* (Contemporary Neurology Series, 3rd ed.). Philadelphia: F. A. Davis.

9. J. Kagan (1989). *Unstable Ideas: Temperament, Cognition, and Self*. Cambridge, MA: Harvard University Press.

CHAPTER 11

1. G. S. Stent (1969). *The Coming of the Golden Age: A View of the End of Progress*. New York: Doubleday.

2. A rich description of this state of affairs can be found in Robert Hughes (1992). *The Culture of Complaint*. New York: Oxford University Press.

3. R. Descartes (1637). *The Philosophical Works of Descartes*, rendered into English by Elizabeth S. Haldane and G. R. T. Ross, vol. 1, page 101. New York: Cambridge University Press (1970).

4. R. Descartes. See note 3 above.

5. R. Cottingham (1992). *A Descartes Dictionary*. Oxford: Blackwell, pg. 36. Plato. *Phaedo* (1971). *The Collected Dialogues of Plato*. E. Hamilton and H. Cairns, eds. Bollingen Series. Pantheon Books. pp. 47–53.

6. See note 3 above.

ENDNOTES FOR POSTSCRIPTUM

1. W. Faulkner (1949). Nobel Prize acceptance speech. The precise context for Faulkner's words was the mounting nuclear threat, but his message is timeless.

2. P. Éluard (1961). Liberté, in G. Pompidou, ed., *Anthologie de la poésie française*. Paris: Hachette.

3. The writings of Jonas Salk and Richard Lewontin (cited above), which these words evoke, contain the optimism and resolve that are indispensable for a comprehensive human biology.

4. See footnote 2, page 290.

5. David Ingvar has also used the term "memories of the future," with precisely the same meaning.

6. Howard Fields (1987). *Pain*. New York: McGraw-Hill Book Co. B. Davis (1994). Behavioral aspects of complex analgesia (to appear).

7. New, less mutilating surgical procedures to manage pain have also been developed since Lima's time. Although prefrontal leucotomy was not as damaging as other so-called psychosurgical procedures, and although it did have the positive result of relieving intractable suffering, it had a negative result as well: the blunting of emotion and feeling, whose long-term consequences are only now being appreciated fully.

Further Reading

The following is a brief list of books pertaining to the topics I have just discussed. This is obviously not a comprehensive list of references. The titles are grouped by general area, but it should be clear that many of them belong in more than one category.

CLASSICAL SOURCES

Darwin, Charles (1872). *The Expression of the Emotions in Man and Animals*. New York: New York Philosophical Library.
Geschwind, N. (1974). *Selected Papers on Language and Brain*. Boston Studies in the Philosophy of Science, Vol. XVI, The Netherlands: D. Reidel Publishing Company.
Hebb, D. O. (1949). *The Organization of Behavior*. New York: Wiley.
James, W. (1890). *The Principles of Psychology*. Volume 1 and 2. New York: Dover Publications (1950).

CURRENT TECHNICAL SOURCES

Churchland, P.S., and T. J. Sejnowski (1992). *The Computational Brain: Models and Methods on the Frontiers of Computational Neuroscience*. Cambridge, MA: Bradford Books, MIT Press.

Damasio, H., and A. R. Damasio (1989). *Lesion Analysis in Neuropsychology*. New York: Oxford University Press.

Damasio, H. (1994). *Human Brain Anatomy in Computerized Images*. New York: Oxford University Press.

Kandel, E. R., J. H. Schwartz, and T. M. Jessell (eds) (1991). *Principles of Neural Science*. 3rd ed. Norwalk, CT: Appleton and Lange.

EMOTION

De Sousa, R. (1991). *The Rationality of Emotion*. Cambridge, MA: MIT Press.

Izard, C. E., J. Kagan, and R. B. Zajonc (1984). *Emotion, Cognition and Behavior*. New York: Cambridge University Press.

Kagan, J. (1989). *Unstable Ideas: Temperament, Cognition, and Self*. Cambridge, MA: Harvard University Press.

Mandler, G. (1984). *Mind and Body: Psychology of Emotion and Stress*. New York: W. W. Norton & Co.

THINKING AND REASONING

Fuster, Joaquin M. (1989). *The Prefrontal Cortex: Anatomy, Physiology, and Neuropsychology of the Frontal Lobe*. 2nd Ed. New York: Raven Press.

Gardner, H. (1983). *Frames of Mind: The Theory of Multiple Intelligences*. New York: Basic Books.

Johnson-Laird, P. N. (1983). *Mental Models*. Cambridge, MA: Harvard University Press.

Pribram, K. H., and A. R. Luria (eds.) (1973). *Psychophysiology of the Frontal Lobe*. New York: Academic Press.

Sutherland, S. (1992). *Irrationality: The Enemy Within*. London: Constable.

FROM PHILOSOPHY OF MIND TO COGNITIVE NEUROSCIENCE

Churchland, P. S. (1986). *Neurophilosophy: Toward A Unified Science of the Mind-Brain*. Bradford Books. Cambridge, MA: MIT Press.

Churchland, P. M. (1984). *Matter and Consciousness*. Cambridge, MA: Bradford Books, MIT Press.

Churchland, P. M. (1994). *The Engine of Reason, The Seat of the Soul: A Philosophical Journey into the Brain*. Cambridge: MIT Press.

Dennett, D. C. (1991). *Consciousness Explained*. New York: Little Brown.

Dudai, Y. (1989). *The Neurobiology of Memory: Concepts, Findings, Trends*. New York: Oxford University Press.

Flanagan, O. (1992). *Consciousness Reconsidered*. Cambridge, MA: MIT Press.

Gazzaniga, M. S., and J. E. Le Doux (1978). *The Integrated Mind*. New York: Plenum Press.

Hinde, R. A. (1990). The Interdependence of the Behavioral Sciences. *Phil. Trans. of the Royal Society*, London, 329, 217–227.

Hubel, D. H. (1987). *Eye, Brain and Vision*. Scientific American Library. Distributed by W. H. Freeman, New York.

Humphrey, N. (1992). *A History of the Mind: Evolution and the Birth of Consciousness*. Norwalk, CT: Simon & Schuster.

Johnson, M. (1987). *The Body in the Mind: The Bodily Basis of Meaning, Imagination, and Reason*. Chicago: University of Chicago Press.

Kosslyn, S. M., and O. Koenig (1992). *Wet Mind: The New Cognitive Neuroscience*. New York: The Free Press.

Lakoff, G. (1987). *Women, Fire, and Dangerous Things: What Categories Reveal About the Mind*. Chicago: University of Chicago Press.

Magnusson, D. (c. 1988). *Individual Development in an Interactional*

Perspective: A Longitudinal Study. Hillsdale, NJ: Erlbaum Associates.

Miller, J. (1983). *States of Mind.* New York: Pantheon Books.

Ornstein, R. (1973). *The Nature of Human Consciousness.* San Francisco: W. H. Freeman.

Rose, S. (1973). *The Conscious Brain.* New York: Knopf.

Rutter, M. and Rutter, M. (1993). *Developing Minds: Challenge and Continuity Across the Lifespan.* New York. Basic Books.

Searle, J. R. (1992). *The Rediscovery of the Mind.* Cambridge, MA: Bradford Books, MIT Press.

Squire, L. R. (1987). *Memory and Brain.* New York: Oxford University Press.

Zeki, S. (1993). *A Vision of the Brain.* Cambridge, MA: Blackwell Scientific Publications.

GENERAL BIOLOGY

Barkow, J. H., L. Cosmides and J. Tooby (eds.) (1992). *The Adapted Mind: Evolutionary Psychology and the Generation of Culture.* New York: Oxford University Press.

Bateson, P. (1991). *The Development and Integration of Behavior: Essays in Honour of Robert Hinde.* New York: Cambridge University Press.

Edelman, G. (1988). *Topobiology.* New York: Basic Books.

Finch, C. E. (1990). *Longevity, Senescence, and the Genome.* Chicago: The University of Chicago Press.

Gould, S. J. (1990). *The Individual in Darwin's World.* Edinburgh, Scotland: Edinburgh University Press.

Jacob, F. (1982). *The Possible and the Actual.* New York: Pantheon Books.

Kauffman, S. A. (1993). *The Origins of Order: Self-Organization and Selection in Evolution.* New York: Oxford University Press.

Lewontin, R. C. (1991). *Biology as Ideology: The Doctrine of DNA.* New York: Harper Perennial.

Medawar, P. B., and J. S. Medawar (1983). *Aristotle to Zoos: A*

Philosophical Dictionary of Biology. Cambridge, MA: Harvard University Press.

Purves, D. (1988). *Body and Brain: A Trophic Theory of Neural Connections.* Cambridge, MA: Harvard University Press.

Salk, J. (1973). *Survival of the Wisest.* New York: Harper Row.

Salk, J. (1985). *The Anatomy of Reality.* New York: Praeger.

Stent, G. S. (ed.) (1978). *Morality as a Biological Phenomenon.* Berkeley: University of California Press.

THEORETICAL NEUROBIOLOGY

Changeux, J.-P. (1985). *Neuronal Man: The Biology of Mind.* L. Garey, trans., New York: Pantheon.

Crick, F. (1994). *The Astonishing Hypothesis: The Scientific Search for the Soul.* New York: Charles Scribner's Sons.

Edelman, G. M. (1992). *Bright Air, Brilliant Fire.* New York: Basic Books.

Koch, C., and J. L. Davis (eds.) (1994). *Large-Scale Neuronal Theories of the Brain.* Cambridge: Bradford Books, MIT Press.

OF GENERAL INTEREST

Blakemore, C. (1988). *The Mind Machine.* New York: BBC Books.

Johnson, G. (1991). *In the Palaces of Memory.* New York: Knopf.

Ornstein, R., and P. Ehrlich (1989). *New World New Mind: Moving Toward Conscious Evolution.* Norwalk, CT: Simon and Schuster.

Restak, R. M. (1988). *The Mind.* New York: Bantam Books.

Scientific American (1992). Special issue on "Mind and Brain."

Acknowledgments

DURING THE preparation of the manuscript I was fortunate to have the advice of several colleagues who read the material and offered suggestions. They were Ralph Adolphs, Ursula Bellugi, Patricia Churchland, Paul Churchland, Francis Crick, Victoria Fromkin, Edward Klima, Frederick Nahm, Charles Rockland, Kathleen Rockland, Daniel Tranel, Gary Van Hoesen, Jonathan Winson, Steven Anderson, Richard Caplan and Arthur Benton. I learned immensely from the friendly debates their comments often prompted, especially when, as sometimes was the case, no agreement was possible. I thank them all for their gift of time, knowledge, and wisdom, although no words are enough to recognize Ralph, Dan, Mrs. Lundy, and Charles for the patience with which they read different versions of several chapters and helped me improve them.

The experience about which I write has been accumulated over a period of about twenty-five years, seventeen of which have been spent at The University of Iowa. I am grateful to my colleagues in the Department of Neurology, especially to the members of the Division of Cognitive Neuroscience (Hanna Damasio, Daniel Tranel, Gary Van Hoesen, Arthur Benton, Kathleen Rockland, Matthew Rizzo,

Thomas Grabowski, Steven Anderson, Ralph Adolphs, Antoine Bechara, Robert Jones, Joseph Barrash, Julie Fiez, Ekaterin Semen-
deferi, Ching-Chiang Chu, Joan Brandt, and Mark Nawrot), for
what they have taught me through the years, and for the spirit and
expertise with which they helped create a unique environment for
the investigation of brain and mind. I am no less grateful to the
neurological patients who have been studied in my unit (and now
number over 1,800), for the opportunity to understand their
problems. I hope the discussions in this book will help them and
their families understand the problems they face. I hope, in particular, that the book will help them explain to others why, on occasion,
they behave as they do.

I wish I could thank John Harlow for the documents he left us on
Phineas Gage. The opening chapters of this book rest on those
documents. In light of our current knowledge, they permit a number
of interesting inferences and conjectures, but they are not the source
of my description of Mr. Adams, or of the weather on the day of the
accident, which are pure literary license.

Betty Redeker prepared the manuscript with the dedication, pro-
fessionalism, and sense of humor that characterize her work. Jon
Spradling and Denise Krutzfeldt helped me with bibliographical
searches with their usual proficiency. Timothy Meyer was the expert copyeditor.

This book would not have been written without the profound
influence and expert guidance of two friends, Michael Carlisle and
Jane Isay, whose enthusiam and loyalty are invaluable.

Hanna Damasio's ideas, findings, criticisms, suggestions, and in-
spiration are an integral part of this book. I would not even try to
thank her for her contributions.

Index

Also available from Vintage

ANTONIO DAMASIO

The Feeling of
What Happens

'A tour de force...a monumental book...a gem of a work'
Sunday Times

One of the world's leading experts on the neurophysiology of
emotions, Professor Damasio shows how our consciousness,
our sense of being, arose out of the development of emotion. At
its core, human consciousness is the consciousness of the feeling, experiencing self, the 'very thought' of oneself. Brilliantly
wide-ranging in his scope, Damasio illustrates his thesis with
fascinating and illuminating neurological case studies that are
both stimulating and provocative.

'There are few more interesting subjects in the world than the
source and workings of consciousness and rarely has there been
an author better qualified to explain them'
Guardian

'A breathtaking and original perspective...stunning...com-
pelling and convincing...written in a language that manages
simultaneously to be sturdily hard-headed and gloriously
poetic: a gem'
Sunday Times

VINTAGE BOOKS
London

Also available from Vintage

ANTONIO DAMASIO

Looking for Spinoza

'Big claims, well made: it is a rare pleasure to pick up such a rigorous and readable book about scientific advance that is so firmly anchored in philosophical history'
Time Out

Joy, sorrow, jealousy and awe – these and other feelings are the stuff of our daily lives. Presumed to be too private for science to explain and not to be essential for comprehending human rationality and understanding, they have largely been ignored. But not by the great seventeenth-century Dutch philosopher Spinoza. And not by Antonio Damasio. In this book Damasio draws on his innovative research and on his experience with neurological patients to examine how feelings and the emotions that underlie them support the governance of human affairs.

'Damasio has important things to say'
New Statesman

'Exceptionally engaging and profoundly gratifying'
Nature

VINTAGE BOOKS
London

www.vintage-books.co.uk